Lecture Notes in Mathematics

continuation on page 203

Lecture Notes in Mathematics

Edited by A. Dold and B. Eckmann

797

Sirpa Mäki

The Determination of
Units in
Real Cyclic Sextic Fields

Springer-Verlag
Berlin Heidelberg GmbH 1980

Author

Sirpa Mäki
Department of Mathematics
University of Turku
20500 Turku 50
Finland

AMS Subject Classifications (1980): 12-04, 12 A 35, 12 A 45, 12 A 50

ISBN 978-3-540-09984-0   ISBN 978-3-540-39242-2 (eBook)
DOI 10.1007/978-3-540-39242-2

© Springer-Verlag Berlin Heidelberg 1980
Originally published by Springer-Verlag Berlin Heidelberg New York in 1980

2141/3140-543210

Contents

1. Introduction

In the preface to his monograph [14] Hasse expresses a wish to
describe the structure of absolutely Abelian number fields in such a
way that one could move in them as freely as in quadratic fields and he
stresses the vital importance of numerical tables and examples for the
carrying out of this programme. Such a table should at least contain
the most important numerical characteristics of each field, i.e. a system of fundamental units and the class number. It is well known that
in the absolutely Abelian case the latter is closely connected with the
former. Tables of this kind had earlier been computed for real quadratic fields by Ince [16], and for real cyclic cubic and quartic fields
by Hasse [13]. More recently, with the aid of a computer, further calculations have been made by Cohn and Gorn [5] and M.N. Gras [9] in the
cubic case and by M.N. Gras [10], [11] in the quartic case.

In spite of the existence of rather clear-cut and far developed
underlying general principles (Hasse [13], [14], Leopoldt [18], [19],
G. and M.N. Gras [7]), the actual computation involves many theoretical
and practical problems for each particular field type. Because of their
rapidly increasing complexity, it will probably not be possible to treat
fields of a considerably high degree. In the present work, which may
be regarded as a continuation of the works of M.N. Gras, we shall deal
with real cyclic fields of degree six. There are five fundamental units,
of which three are known, i.e. those belonging to the proper subfields.
To determine the missing two units we start off from the so-called cyclotomic (or circular) unit which is calculable from a definite expres-

sion. This unit, together with its conjugates, and the units of the proper subfields generate a subgroup of finite index in the whole unit group, and it is in principle relatively easy to obtain the whole group from this subgroup. The real cyclic sextic case is thus sufficiently simple to enable one to compute large tables, but on the other hand the existence of subfields of different types hopefully renders it sufficiently complex for it to possess several features appearing in the general case.

A real cyclic sextic field K_6 has unique proper subfields of degree 2 and 3 denoted by K_2 and K_3. In Section 2 we gather together some basic results about the cubic subfield K_3 which are needed in the sequel. Later on an important role is played by the formula (15), due essentially to Hasse [13, p. 15], which leads to useful estimates. Section 3 first contains some generalities concerning the field K_6. For example, we give a necessary condition for a number of K_6 to be an algebraic integer. Further we prove some results (Theorems 2 and 3) to the effect that under certain conditions a unit of K_6 cannot be a power in K_6 nontrivially. The bulk of Section 3 is devoted to the study of the structure of the unit group U_6 of the field K_6. The group U_6 has the unit groups U_2 and U_3 of the subfields K_2 and K_3 as subgroups. U_2 is generated by -1 and the fundamental unit μ, and U_3 is generated by -1, a fundamental unit τ and one of its conjugates τ'. An element of U_6 is called a relative unit if its relative norms to the subfields K_2 and K_3 are both ± 1. The relative units of K_6 form a subgroup U_R of U_6. In the classic way we define ξ to be the number λ in [14, p. 21] and let η denote the cyclotomic unit of K_6. We take ξ_A to be ξ or η according as $\xi \in K_6$ or $\xi \notin K_6$. It is seen that ξ_A is a unit of K_6. We show that the subgroup U_6^* of U_6 generated by $U_2 U_3 U_R$ and the conjugates of ξ_A is of index 12 at most. Therefore to obtain U_6 from U_6^* it is enough to decide the existence or nonexistence of certain supplementary units, denoted by ξ_B and ξ_C, satisfying given equations. At the end of Section 3 we give a simple proof for the fact that the system $\{\mu,\tau,\tau'\}$ can be completed to a system of fundamental units of K_6.

The question of determining U_6 is thus essentially reduced to questions about U_R, and in Section 4 we shall be concerned with the structure of U_R. In the actual computation the most important practical problem is the question whether a given element of U_R is a nontrivial power in U_R or not. In solving this a very useful tool is the mean square modulus function $\mathcal{M}$, defined in (50). Its properties have been elegantly exploited by Leopoldt [19], Cassels [3], and Loxton [21]. The formula (15), mentioned above, enables one to find a very efficient lower bound for the function $\mathcal{M}$ in $U_R \smallsetminus \{\pm 1\}$ (Theorem 10). This gives us the good upper estimate K_{max}, defined in (73), for the possible exponent. We shall further derive some useful monotonicity results concerning the function $\mathcal{M}$ in $U_R \smallsetminus \{\pm 1\}$. Define ξ_R so that $\mathcal{M}(\xi_R)$ is the least value of $\mathcal{M}$ in $U_R \smallsetminus \{\pm 1\}$. We give a proof of a result of Leopoldt [19] to the effect that ξ_R is a generating relative unit, i.e. ξ_R together with its conjugates and -1 generates U_R.

At the end of Section 4 we introduce a candidate ξ_o for a generating relative unit. The relative unit ξ_o is formed from ξ_A in a natural way. Using the upper bound K_{max} and the monotonicity properties of the function $\mathcal{M}$ one can find out whether or not ξ_o is a generating relative unit, and determine ξ_R.

In Bergström's product formula the number ξ is represented by means of Gaussian sums. Section 5 is somewhat expanded from Hasse's work [13] concerning this formula. In Section 6 real cyclic sextic fields are divided into ten classes. The first class consists of those fields K_6 having a decomposable conductor f_6 in the terminology of Hasse [13] (in German zerlegbar). By this we mean that a generating character of K_6 is decomposable into a product of two nonprincipal even characters having relatively prime conductors. We have $\xi \in K_6$ if and only if f_6 is decomposable. On p.51 we give a set of conditions each one of which is equivalent to the decomposability of f_6. The fields for which f_6 is not decomposable are further divided into nine classes depending on the prime factorizations of the conductors of K_2 and K_3. Bergström´s product formula is then developed in each of these ten cases separately. Section 7 contains formulas for computing η from Bergström´s product formula, and a number of other useful formulas.

In Section 8 we show that the class number h_6 of K_6 is of the form $h_6 = h_2 h_3 h_R$ where h_2 and h_3 are the class numbers of K_2 and K_3, and h_R is a natural number called the relative class number of K_6. We also give a formula expressing h_R as a group-index.

The signature rank Sr of U_6 is defined as the dimension of the image of the natural signature homomorphism from U_6 into the additive group of a vector space of dimension six over GF(2). Section 9 contains results needed in the practical computation of Sr. E.g., we prove that the difference between Sr and the signature rank of the subgroup $U_2 U_3$ is always 2 or 0, and we show how Sr can be determined without computing any signatures if τ is not totally positive.

In Section 10 we give a rather exhaustive technical description of our computer program. Our choice of τ among its conjugates is explained on p. 64.

We have investigated all real cyclic sextic fields K_6 with conductor $f_6 \leqslant 2021$, and obtained a complete numerical result in each case with 12 exceptions. The reason for the latter failure is absence of information concerning K_3 or the appearance of too large numbers which we have been unable to handle. The smallest value of f_6 for these exceptions is 997. We have discovered altogether 130 fields with $h_R > 1$, the largest observed value of h_R being 16.

In the bibliography only those works are listed which are of relevance to our particular study. Extensive bibliographies concerning related topics will be found, for example, in Masley [22], Shanks [24], and Zimmer [28].

In the list of terminology and notations we have tried to put together all the most important symbols and notions used in this work. As regards any unexplained notation appearing in the text, one should consult this list.

This work has been supported financially by the Academy of Finland.

2. Real cyclic cubic fields

Let K_3 be a real cyclic cubic field. Its conductor f_3 is of the form

(1)
$$f_3 = \begin{cases} p_0 p_1 \cdots p_n & \text{if } 3 \nmid f_3 \\ 9 p_1 \cdots p_n & \text{if } 3 \mid f_3 \end{cases}$$

where p_i is a prime $\equiv 1 \bmod 3$ $(i = 0,1,\ldots,n)$ and $p_i \neq p_j$ $(i \neq j)$ [13,p.10]. There exist uniquely determined rational integers a and b such that

(2)
$$f_3 = (a^2 + 3b^2)/4,$$

(3)
$$\begin{cases} a \equiv 2, \ b \equiv 0 \bmod 3 \text{ and } b > 0 \text{ if } 3 \nmid f_3 \\ \begin{cases} a = 3a_0, \ b = 3b_0 \\ a_0 \equiv 2, \ b_0 \not\equiv 0 \bmod 3 \text{ and } b_0 > 0 \text{ if } 3 \mid f_3. \end{cases} \end{cases}$$

Put
$$\phi = (a + b\sqrt{-3})/2.$$

There is a one-to-one correspondence between such pairs a,b and real cyclic cubic fields having conductor f_3. (Note that in [8],[9] M.N. Gras uses the notation where $f_3 = (a^2 + 27b^2)/4$, $b > 0$, $a \equiv 1 \bmod 3$ if $3 \nmid f_3$ and $a = 3a'$, $a' \equiv 1 \bmod 3$ if $3 \mid f_3$. In this paper we use the notation of Hasse [13].)

The discriminant of K_3 is

(4)
$$d_3 = f_3^2$$

[13. p. 13].

Put

(5)
$$\theta = (-1)^n {\sum_x}' \zeta_{f_3}^x$$

where ${\sum_x}'$ denotes a summation over all x mod f_3 such that $\chi(x) = 1$ for a generating character χ of K_3, and $\zeta_{f_3} = e^{2\pi i/f_3}$. Then $K_3 = \mathbb{Q}(\theta)$. Let $u \in \mathbb{Z}$ be such that g.c.d.$(f_3, u) = 1$ and $\chi(u) \neq 1$. Then the other conjugates of θ are

$$(-1)^n {\sum_x}' \zeta_{f_3}^{ux} \;, \quad (-1)^n {\sum_x}' \zeta_{f_3}^{u^2 x} \;.$$

We denote one of these by θ' and the other by θ''. We shall fix the notation later. The numbers $\theta, \theta', \theta''$ are algebraic integers in K_3. In the following we shall derive the minimal polynomial of θ and we shall give the formulas for θ' and θ'' in terms of θ.

Case 1. $3 \nmid f_3$. In this case

$$\theta + \theta' + \theta'' = (-1)^n \sum_{\substack{x=1 \\ (x,f_3)=1}}^{f_3} \zeta_{f_3}^x = -1.$$

The character $\chi(x) = \left(\dfrac{x}{\phi}\right)_3$ is a generating character of K_3 [13, p. 12]. (Cf. p. 37 where χ is denoted by χ_3'.) Put $\rho = (-1 + \sqrt{-3})/2$. The Gaussian sum $\tau(\chi)$ for the character χ is

$$\tau(\chi) = \sum_{\substack{x=1 \\ (x,f_3)=1}}^{f_3} \chi(x) \zeta_{f_3}^x \;.$$

Hence either $\tau(\chi) = (-1)^n(\theta + \rho\theta' + \rho^2\theta'')$ or $\tau(\chi) = (-1)^n(\theta + \rho^2\theta' + \rho\theta'')$ so that

$$f_3 = \tau(\chi)\overline{\tau(\chi)} = (\theta + \theta' + \theta'')^2 - 3(\theta\theta' + \theta'\theta'' + \theta''\theta)$$

$$= 1 - 3(\theta\theta' + \theta'\theta'' + \theta''\theta).$$

Thus $\theta\theta' + \theta'\theta'' + \theta''\theta = (1 - f_3)/3$. Since

$$\tau(\chi)^3 = (-1)^{n+1} f_3 \frac{a + b\sqrt{-3}}{2}$$

[13, p. 13], we have for k = 1 or 2

$$-f_3 \frac{a + b\sqrt{-3}}{2} = (\theta + \theta' + \theta'')^3 + 3(\rho^k - 1)(\theta^2\theta' + \theta'^2\theta'' + \theta''^2\theta)$$

$$+ 3(\rho^{2k} - 1)(\theta\theta'^2 + \theta'\theta''^2 + \theta''\theta^2).$$

Taking real parts we obtain

$$\frac{f_3 a - 2}{9} = \theta^2\theta' + \theta'^2\theta'' + \theta''^2\theta + \theta\theta'^2 + \theta'\theta''^2 + \theta''\theta^2$$

from which it follows that

$$\theta\theta'\theta'' = \frac{1}{3}\{(\theta + \theta' + \theta'')(\theta\theta' + \theta'\theta'' + \theta''\theta)$$

$$- (\theta^2\theta' + \theta'^2\theta'' + \theta''^2\theta + \theta\theta'^2 + \theta'\theta''^2 + \theta''\theta^2)\}$$

$$= \frac{f_3(3 - a) - 1}{27} .$$

Hence the minimal polynomial of θ is

(6) $\qquad \mathrm{Irr}(\theta, \mathbb{Q}) = x^3 + x^2 + \dfrac{1 - f_3}{3} x + \dfrac{f_3(a - 3) + 1}{27} .$

The other conjugates of θ are

(7) $\qquad \theta' = \dfrac{-4f_3 + a + 2}{6b} - \dfrac{1}{2} + (\dfrac{a + 4}{2b} - \dfrac{1}{2})\theta + \dfrac{3}{b} \theta^2,$

(8) $\qquad \theta'' = \dfrac{4f_3 - a - 2}{6b} - \dfrac{1}{2} + (\dfrac{-a - 4}{2b} - \dfrac{1}{2})\theta - \dfrac{3}{b} \theta^2$

[4, p. 121].

Case 2. $3 | f_3$. In this case $\theta + \theta' + \theta'' = 0$. Define $\alpha = \pm 1$ by $\phi \equiv 3\rho^\alpha$ mod 9. Then $\chi(x) = \left(\dfrac{\rho}{x}\right)_3^\alpha \left(\dfrac{x}{\phi}\right)_3$ is a generating character of K_3. (Cf. p. 37 where χ is denoted by χ_3'. We define formally $\left(\dfrac{x}{3\rho^\alpha}\right)_3 = \left(\dfrac{x}{3}\right)_3$ = 1 for all x such that $3 \nmid x$.) As in Case 1 we obtain $\theta\theta' + \theta'\theta'' + \theta''\theta = -f_3/3$. From the equation

$$\tau(\chi)^3 = (-1)^n f_3 \frac{a + b\sqrt{-3}}{2}$$

[13, p. 13] it follows as in Case 1 that $\theta\theta'\theta'' = f_3 a/27$. Hence the minimal polynomial of θ is

(9)
$$\text{Irr}(\theta, \mathbb{Q}) = x^3 - \frac{f_3}{3} x - \frac{f_3 a}{27} .$$

The other conjugates of θ are

(10)
$$\theta' = -\frac{2f_3}{3b} + (-\frac{a}{2b} - \frac{1}{2})\theta + \frac{3}{b}\theta^2 ,$$

(11)
$$\theta'' = \frac{2f_3}{3b} + (\frac{a}{2b} - \frac{1}{2})\theta - \frac{3}{b}\theta^2$$

[4, p. 121].

In both cases we fix the conjugates θ', θ'' so that θ' is always that conjugate in which θ^2 has the coefficient $\frac{3}{b}$ and correspondingly θ'' is that conjugate in which the coefficient of θ^2 is $-\frac{3}{b}$. Let an integer u be such that

$$\theta' = (-1)^n \sum_x {}' \zeta_{f_3}^{ux} .$$

Let χ_3 denote that generating character of K_3 for which $\chi_3(u) = \rho$.

In both cases $\{1, \theta, \theta'\}$ is an integral basis of K_3, because

(12)
$$\begin{vmatrix} 1 & \theta & \theta' \\ 1 & \theta' & \theta'' \\ 1 & \theta'' & \theta \end{vmatrix} = -(\theta + \theta' + \theta'')^2 + 3(\theta\theta' + \theta'\theta'' + \theta''\theta) = -f_3 .$$

Later in this paper we shall need formulas from which we obtain the coordinates of the product of two elements of K_3. Let

$$\alpha = x_0 + x_1\theta + x_2\theta' ,$$

$$\beta = y_0 + y_1\theta + y_2\theta' .$$

Case 1. $3 \nmid f_3$. From (6), (7) and (8) we obtain

$$\theta^2 = \frac{4f_3 - a + 3b - 2}{18} + \frac{-a + b - 4}{6}\theta + \frac{b}{3}\theta' ,$$

$$\theta\theta' = \frac{-f_3 + a - 1}{9} + \frac{a + b - 2}{6}\theta + \frac{a - b - 2}{6}\theta' ,$$

$$\theta'^2 = \frac{4f_3 - a - 3b - 2}{18} - \frac{b}{3}\theta - \frac{a+b+4}{6}\theta' \;.$$

By means of these equations we obtain

$$(13) \quad \alpha\beta = x_0y_0 + \frac{4f_3 - a + 3b - 2}{18} x_1y_1 + \frac{-f_3 + a - 1}{9}(x_1y_2 + x_2y_1)$$

$$+ \frac{4f_3 - a - 3b - 2}{18} x_2y_2$$

$$+ \{x_0y_1 + x_1y_0 + \frac{-a+b-4}{6} x_1y_1 + \frac{a+b-2}{6}(x_1y_2 + x_2y_1)$$

$$- \frac{b}{3} x_2y_2\}\theta$$

$$+ \{x_0y_2 + \frac{b}{3} x_1y_1 + \frac{a-b-2}{6}(x_1y_2 + x_2y_1) + x_2y_0$$

$$- \frac{a+b+4}{6} x_2y_2\}\theta' \;.$$

<u>Case 2.</u> $3|f_3$. From (9), (10) and (11) we have

$$\theta^2 = \frac{2f_3}{9} + \frac{a+b}{6}\theta + \frac{b}{3}\theta' \;,$$

$$\theta\theta' = -\frac{f_3}{9} - \frac{a-b}{6}\theta - \frac{a+b}{6}\theta' \;,$$

$$\theta'^2 = \frac{2f_3}{9} - \frac{b}{3}\theta + \frac{a-b}{6}\theta' \;,$$

from which we obtain

$$(14) \quad \alpha\beta = x_0y_0 + \frac{2f_3}{9} x_1y_1 - \frac{f_3}{9}(x_1y_2 + x_2y_1) + \frac{2f_3}{9} x_2y_2$$

$$+ \{x_0y_1 + x_1y_0 + \frac{a+b}{6} x_1y_1 - \frac{a-b}{6}(x_1y_2 + x_2y_1) - \frac{b}{3} x_2y_2\}\theta$$

$$+ \{x_0y_2 + \frac{b}{3} x_1y_1 - \frac{a+b}{6}(x_1y_2 + x_2y_1) + x_2y_0 + \frac{a-b}{6} x_2y_2\}\theta' .$$

We shall also need $S_{3/1}(\alpha^2)$ i.e. the trace of α^2 from K_3 to $\mathbb{Q}$. Now

$$S_{3/1}(\alpha^2) = 3x_0^2 + (x_1^2 + x_2^2)S_{3/1}(\theta^2)$$

$$+ 2(x_0x_1 + x_0x_2)S_{3/1}(\theta) + 2x_1x_2S_{3/1}(\theta\theta').$$

In the case $3 \nmid f_3$ we have from p. 7 that $S_{3/1}(\theta) = -1$, $S_{3/1}(\theta\theta') = (1 - f_3)/3$ and $S_{3/1}(\theta^2) = (S_{3/1}(\theta))^2 - 2S_{3/1}(\theta\theta') = (2f_3 + 1)/3$. So

$$S_{3/1}(\alpha^2) = \frac{1}{3}(3x_0 - x_1 - x_2)^2 + \frac{2f_3}{3}(x_1^2 + x_2^2 - x_1x_2).$$

In the case $3 \mid f_3$ we have from p. 8 that $S_{3/1}(\theta) = 0$, $S_{3/1}(\theta\theta') = -f_3/3$ and $S_{3/1}(\theta^2) = 2f_3/3$. Hence

$$S_{3/1}(\alpha^2) = 3x_0^2 + \frac{2f_3}{3}(x_1^2 + x_2^2 - x_1x_2).$$

Thus in both cases

(15) $$S_{3/1}(\alpha^2) = \frac{1}{3}S_{3/1}(\alpha)^2 + \frac{2f_3}{3}(x_1^2 + x_2^2 - x_1x_2).$$

According to Hasse [13, p. 20] there exists in K_3 a unit τ such that τ together with one of its conjugates generates the group of units $\varepsilon \in K_3$ each having norm $N_{3/1}(\varepsilon) = 1$. The fundamental unit τ can be found by taking such a unit $\varepsilon \in K_3 \setminus \{1\}$ with $N_{3/1}(\varepsilon) = 1$ for which $S_{3/1}(\varepsilon^2)$ is the least possible. The fundamental unit τ is uniquely determined except for taking conjugates and inverses. If $\tau_1 \neq \tau$ is any conjugate of τ, then the group U_3 of units of K_3 is generated by -1, τ, τ_1.

3. Real cyclic sextic fields

Let G be the Galois group of a real cyclic sextic field K_6. G has exactly two nontrivial subgroups, namely those of order 2 and 3. Thus the field K_6 has exactly two nontrivial subfields: a real cyclic cubic field K_3 and a real quadratic field K_2. We have $K_3 = Q(\theta)$ where θ is the number defined in (5), and $K_2 = Q(\sqrt{m})$ where $m > 1$ is a square-free integer. Hence $K_6 = Q(\theta, \sqrt{m})$.

Let an odd integer s be such that the automorphism σ of K_6 induced by the mapping $\zeta_{f_6} \mapsto \zeta_{f_6}^s$, where f_6 is the conductor of K_6, satisfies the conditions

$$\sigma(\theta) = \theta', \quad \sigma(\theta') = \theta'', \quad \sigma(\sqrt{m}) = -\sqrt{m}$$

using the notation introduced on p. 9 . Let Φ_{f_6} be the multiplicative group of prime residue classes mod f_6, and let H be the subgroup of Φ_{f_6} consisting of those elements $x + f_6 Z$ for which the automorphisms $\zeta_{f_6} \mapsto \zeta_{f_6}^x$ of the field $Q(\zeta_{f_6})$ keep K_6 elementwise fixed. Then $\Phi_{f_6}/H = \{H, sH, \ldots, s^5 H\}$. We denote the conjugates of a number γ in K_6 in the following way:

(16)
$$\begin{cases} \gamma' = \sigma(\gamma), \quad \gamma'' = \sigma^2(\gamma), \quad \gamma''' = \sigma^3(\gamma), \\ \gamma^{iv} = \sigma^4(\gamma), \quad \gamma^v = \sigma^5(\gamma). \end{cases}$$

Sometimes we also use the notation

(17)
$$\gamma^{(i)} = \sigma^i(\gamma) \quad (i \in Z).$$

If K_6 is a subfield of a cyclotomic field $Q(\zeta_k)$ then also $Q(\zeta_{f_2})$

and $\mathbb{Q}(\zeta_{f_3})$ are contained in $\mathbb{Q}(\zeta_k)$. Hence

(18) $$f_6 = \text{l.c.m.}(f_2, f_3).$$

The conductor f_2 of $\mathbb{Q}(\sqrt{m})$ is

$$f_2 = \begin{cases} 4m \text{ when } m \equiv 2,3 \bmod 4 \\ m \text{ when } m \equiv 1 \bmod 4. \end{cases}$$

The conductor f_3 of $\mathbb{Q}(\theta)$ we recall from (1).

The characters of K_6 are the principal character χ_1, the quadratic character χ_2 of K_2, the generating characters χ_3 and $\overline{\chi}_3$ of K_3 and the generating characters $\chi_6 = \chi_2\chi_3$ and $\overline{\chi}_6 = \chi_2\overline{\chi}_3$ of K_6. The conductor of the character χ_n and $\overline{\chi}_n$ is f_n.

The discriminant d_6 of the field K_6 is

(19) $$d_6 = f_6{}^2 f_3{}^2 f_2$$

[14, p. 8].

Let $\mathcal{O}_n$ denote the ring of integers of K_n. The next theorem gives a necessary condition for a number $\gamma \in K_6$ to belong to $\mathcal{O}_6$.

<u>Theorem 1.</u> If $\gamma \in \mathcal{O}_6$ then γ is of the form

(20) $$\gamma = \frac{1}{2}(x_0 + x_1\theta + x_2\theta') + \frac{1}{2f_*}(y_0 + y_1\theta + y_2\theta')\sqrt{m}$$

where $f_* = \text{g.c.d.}(f_2, f_3)$ and $\frac{1}{2}x_i + \frac{1}{2}y_i\sqrt{m} \in \mathcal{O}_2$ ($i = 0,1,2$).

<u>Proof.</u> Let $\gamma = a_0 + a_1\theta + a_2\theta' + (b_0 + b_1\theta + b_2\theta')\sqrt{m}$, where $a_i, b_i \in \mathbb{Q}$ ($i = 0,1,2$), be a number of $\mathcal{O}_6$. Then $\gamma + \gamma''' = 2(a_0 + a_1\theta + a_2\theta')$ belongs to $\mathcal{O}_3$. We recall from p. 9 that $\{1,\theta,\theta'\}$ is an integral basis of K_3. So $a_i = x_i/2$ where $x_i \in \mathbb{Z}$ ($i = 0,1,2$). Also $\sqrt{m}(\gamma - \gamma''')$ $= 2m(b_0 + b_1\theta + b_2\theta')$ is an algebraic integer. Hence $b_i = z_i/(2m)$ where $z_i \in \mathbb{Z}$ ($i = 0,1,2$). Put $\lambda_i = \frac{1}{2}x_i + \frac{1}{2m}z_i\sqrt{m}$. Now we have

$$\begin{cases} \gamma & = \lambda_0 + \lambda_1\theta + \lambda_2\theta' \\ \gamma^{iv} & = \lambda_0 + \lambda_1\theta' + \lambda_2\theta'' \\ \gamma'' & = \lambda_0 + \lambda_1\theta'' + \lambda_2\theta \ . \end{cases}$$

The determinant of this system of linear equations is $-f_3$ by (12). Thus the numbers $f_3\lambda_i$ ($i = 0,1,2$) are algebraic integers. Since $f_3\lambda_i \in \mathcal{O}_2$ ($i = 0,1,2$), the numbers $f_3 z_i/m$ ($i = 0,1,2$) are rational integers. Since f_3 is odd, $(m,f_3) = (f_2,f_3) = f_*$. Thus $z_i/m = y_i/f_*$ where $y_i \in \mathbb{Z}$ ($i = 0,1,2$). Also the numbers $\frac{1}{2}x_i + \frac{1}{2}y_i\sqrt{m} \in \mathcal{O}_2$, because f_3 is odd and $f_3\lambda_i$ is integral. □

According to Dirichlet's theorem on units in K_6 there are 5 fundamental units, which together with -1 generate the multiplicative group U_6 of units of K_6. The groups U_3 and U_2 of units of K_3 and K_2 respectively are subgroups of U_6. Let μ denote the fundamental unit of K_2. The group U_2 is generated by -1 and μ. The group U_3 is generated by -1, the fundamental unit τ and $\tau' = \sigma(\tau)$.

Units ε of K_6 for which $N_{6/3}(\varepsilon) = \pm 1$ and $N_{6/2}(\varepsilon) = \pm 1$ are called relative units. Let U_R denote the group of such units, i.e.

(21) $$U_R = \{\varepsilon \in U_6 \mid N_{6/3}(\varepsilon) = \pm 1, \ N_{6/2}(\varepsilon) = \pm 1\}.$$

If $\varepsilon \in U_R$, then $N_{6/1}(\varepsilon) = N_{2/1}(N_{6/2}(\varepsilon)) = N_{2/1}(\pm 1) = 1$. On the other hand $N_{6/1}(\varepsilon) = N_{3/1}(N_{6/3}(\varepsilon)) = (N_{6/3}(\varepsilon))^3$ so that for $\varepsilon \in U_R$

(22) $$N_{6/3}(\varepsilon) = 1.$$

Using the notation (17) we obtain from (22) the equation

(23) $$N_{6/3}(\varepsilon^{(i)}) = \varepsilon^{(i)}\varepsilon^{(i+3)} = 1$$

for all integers i. From (21) we also have

(24) $$N_{6/2}(\varepsilon^{(i)}) = \varepsilon^{(i)}\varepsilon^{(i+2)}\varepsilon^{(i+4)} = \pm 1.$$

From the equations (23) and (24) we obtain a useful formula

(25) $$\varepsilon^{(i+1)} = \pm\varepsilon^{(i)}\varepsilon^{(i+2)}.$$

Using (25) one can express each conjugate $\varepsilon^{(i)}$ in terms of ε, ε' if $\varepsilon \in U_R$. The same holds for any $\varepsilon \in U_6$ modulo the subgroup $U_2 U_3$. These expressions are often needed in the sequel. In Section 4 we shall prove that in K_6 there exists a generating relative unit ξ_R such that

(26) $$U_R = \{\pm\xi_R^{\,k}\xi_R'^{\,l} \mid k,l \in \mathbb{Z}\}.$$

Furthermore every $\varepsilon \in U_R$ has a unique representation in the form $\varepsilon = (-1)^\nu \xi_R^{\,k}\xi_R'^{\,l}$ where $\nu \in \{0,1\}$ and the numbers k, l are integers.

The next two theorems are in some cases suitable when we want to find out if a unit of K_6 is a power of an element in K_6.

Theorem 2. Let ε be a unit of K_6 such that $\varepsilon^k \in K_n$ where k is a positive integer and n = 2 or 3. Then $\varepsilon \in K_n$.

Proof. Suppose that $\varepsilon \in K_6 \smallsetminus K_n$. Then $\varepsilon^{(n)} = \sigma^n(\varepsilon) \neq \varepsilon$. Since $(\varepsilon/\varepsilon^{(n)})^k = 1$, k must be even and $\varepsilon^{(n)} = -\varepsilon$. Put $\delta = \prod_{i=0}^{n-1}\varepsilon^{(i)}$. We thus have $\sigma(\delta) = -\delta$. On the other hand $\delta^k = \prod_{i=0}^{n-1}\sigma^i(\varepsilon^k) \in \mathbb{Q}$ so that $\delta = \pm 1$, a contradiction. $\square$

Theorem 3. Let $\varepsilon \in U_R$ and let $\varepsilon = \alpha + \beta\sqrt{m}$ where $\alpha,\beta \in K_3$. If there exist units $\delta \in U_3$ and $\omega \in U_6$ such that $\varepsilon = \delta\omega^2$, then

$$2\alpha \equiv \pm 2 \bmod f_2'$$

where $f_2' = f_2/(f_2,f_3)$.

Proof. Let $\omega = \gamma + \lambda\sqrt{m}$ where $\gamma,\lambda \in K_3$. Now $\varepsilon = \alpha + \beta\sqrt{m} = \delta(\gamma + \lambda\sqrt{m})^2 = \delta(\gamma^2 + m\lambda^2 + 2\gamma\lambda\sqrt{m})$. Thus

(27) $$\alpha = \delta(\gamma^2 + m\lambda^2).$$

Since $\varepsilon = \delta\omega^2$, $\varepsilon \in U_R$ and $\delta \in U_3$, we have $1 = \delta^2(N_{6/3}(\omega))^2$. Thus

(28) $$N_{6/3}(\omega) = \gamma^2 - m\lambda^2 = \frac{e}{\delta},$$

where e = ± 1. From (27) and (28) we now have

$$2\alpha = 2e + 4m\delta\lambda^2$$

where $4m = f_2$ or $m = f_2$. According to Theorem 1, $2\alpha \in \mathcal{O}_3$. Let p be an odd prime such that $p \mid f_2'$. Since $\omega = \gamma + \lambda\sqrt{m} \in \mathcal{O}_6$, λ is p-integral according to Theorem 1. Hence

(29) $$2\alpha \equiv 2e \bmod p.$$

If $2^\nu \| f_2$ $(\nu > 0)$, then λ is 2-integral. Thus

(30) $$2\alpha \equiv 2e \bmod 2^\nu.$$

From (29) and (30) we have $2\alpha \equiv 2e \bmod f_2'$. □

If $x + f_6\mathbb{Z} \in H$, which is defined on p.12, then also $-x + f_6\mathbb{Z} \in H$, because K_6 is real. From every pair $x + f_6\mathbb{Z}$, $-x + f_6\mathbb{Z}$ choose one residue class and then odd representatives for these residue classes. Let α denote the set of these representatives. The number

(31) $$\xi = \prod_{x \in \alpha} (\zeta_{2f_6}^{x} - \zeta_{2f_6}^{-x})$$

is an integer of $\mathbb{Q}(\zeta_{2f_6})$. Put

(32) $$\xi' = \prod_{x \in \alpha} (\zeta_{2f_6}^{sx} - \zeta_{2f_6}^{-sx}),$$

where s is the number defined on p. 12, and

(33) $$\eta = \frac{\xi}{\xi'}.$$

Then $\eta \in K_6$ and η is the cyclotomic unit of K_6 as defined in [14, p.25]. Depending on the choice of α, ξ may have one of two values of opposite signs. Similarly, the sign of η is affected by the choice of s. Take $\xi_A = \xi$ if $\xi \in K_6$ and $\xi_A = \eta$ if $\xi \notin K_6$. Plainly ξ is a unit if f_6 is not a prime power. On the other hand, we shall see later that f_6 cannot be a prime power in the case $\xi \in K_6$ (cf. the equivalent conditions on p. 51). Thus $\xi_A \in U_6$ and we can write

(34) $$N_{6/3}(\xi_A) = \pm\tau^u\tau'^v,$$

(35)
$$N_{6/2}(\xi_A) = \pm\mu^w$$

for some integers u, v, w. Let U_6^* be the group

(36)
$$U_6^* = <-1,\mu,\tau,\tau',\xi_A,\xi_A',\xi_R,\xi_R'> .$$

From the observations after the equation (25) it is clear that $\sigma(\epsilon) \in U_6^*$ for any $\epsilon \in U_6^*$.

Theorem 4. Let u, v, w be the integers in (34) and (35). Then

$$<-1>N_{6/3}(U_6^*) = \begin{cases} U_3 & \text{if } 2{\not|}u \text{ or } 2{\not|}v \\ <-1,\tau^2,\tau'^2> & \text{if } 2|u \text{ and } 2|v \end{cases}$$

and

$$N_{6/2}(U_6^*) = \begin{cases} U_2 & \text{if } 3{\not|}w \\ <-1,\mu^3> & \text{if } 3|w. \end{cases}$$

Proof. If $2{\not|}u$ or $2{\not|}v$, there exist integers c and d such that $N_{6/3}(\tau^c\tau'^d\xi_A) = \pm\tau^{2c+u}\tau'^{2d+v} = \pm\tau, \pm\tau'$, or $\pm\tau\tau'$. Hence $\tau,\tau' \in <-1>N_{6/3}(U_6^*)$ and $<-1>N_{6/3}(U_6^*) = U_3$. If $2|u$ and $2|v$, then for all $\epsilon \in U_6^*$, $N_{6/3}(\epsilon)$ is of the form $\pm\tau^{2g}\tau'^{2h}$. On the other hand $\tau^2 = N_{6/3}(\tau) \in N_{6/3}(U_6^*)$ and $\tau'^2 = N_{6/3}(\tau') \in N_{6/3}(U_6^*)$, so that $<-1>N_{6/3}(U_6^*) = <-1,\tau^2,\tau'^2>$.

If $3{\not|}w$, there exists an integer e such that $N_{6/2}(\mu^e\xi_A) = \pm\mu^{3e+w} = \pm\mu$ or $\pm\mu^{-1}$. Thus $\mu \in N_{6/2}(U_6^*)$. Since $-1 = N_{6/2}(-1) \in N_{6/2}(U_6^*)$ we have $N_{6/2}(U_6^*) = U_2$. If $3|w$, then for all $\epsilon \in U_6^*$, $N_{6/2}(\epsilon)$ is of the form $\pm\mu^{3k}$. Since $\mu^3 = N_{6/2}(\mu) \in N_{6/2}(U_6^*)$ we have $N_{6/2}(U_6^*) = <-1,\mu^3>$.□

In the next four theorems the structure of U_6 is illustrated. Similar results are contained in Yokoi [27]. There are somewhat analogous considerations in the pure sextic case in Stender [25].

Theorem 5. If $<-1>N_{6/3}(U_6^*) = U_3$ and $N_{6/2}(U_6^*) = U_2$, then $U_6 = U_6^*$.

Proof. Let ϵ be any element of U_6. Since $<-1>N_{6/3}(U_6^*) = U_3$ there exists an $\omega_1 \in U_6^*$ such that $N_{6/3}(\omega_1) = \pm N_{6/3}(\epsilon)$. Since $N_{6/2}(U_6^*) = U_2$ there exists an $\omega_2 \in U_6^*$ such that $N_{6/2}(\omega_2) = N_{6/2}(\epsilon/\omega_1) = \pm\mu^e$, say.

Let $N_{6/3}(\omega_2) = \pm\tau^c\tau'^d$. Then $\varepsilon\omega_1^{-1}\omega_2^2\mu^{-e}\tau^{-c}\tau'^{-d} \in U_R$ whence $\varepsilon \in U_6^*$. $\square$

Theorem 6. Let $<\!-1\!>N_{6/3}(U_6^*) = U_3$ and $N_{6/2}(U_6^*) \neq U_2$.

(i) If either one of the equations

(37) $$x^3 = \mu\xi_R\xi_R' , \quad x^3 = \mu^{-1}\xi_R\xi_R'$$

has a solution $x = \xi_B \in K_6$, then $[U_6 : U_6^*] = 3$,
$U_6 = U_6^* \cup \xi_B U_6^* \cup \xi_B^2 U_6^*$ and $N_{6/2}(U_6) = U_2$.

(ii) If neither of the equations (37) has a solution in K_6, then
$U_6 = U_6^*$ and $[U_2 : N_{6/2}(U_6)] = 3$.

Remark. Both the equations (37) cannot have a solution in K_6. If namely $\omega_1^3 = \mu\xi_R\xi_R'$ and $\omega_2^3 = \mu^{-1}\xi_R\xi_R'$ then $(\omega_1/\omega_2)^3 = \mu^2$. This implies that μ is a third power of an element in K_6, a contradiction according to Theorem 2.

Proof. Let us first suppose that the equation $x^3 = \mu^e\xi_R\xi_R'$ ($e = \pm1$) has a solution $\xi_B \in K_6$. We have $N_{6/2}(\xi_B) = \mu^e$, $N_{6/3}(\xi_B) = \pm1$, and it is easy to see, using Theorem 4 and the argument in the proof of Theorem 5, that (i) is true.

Secondly we shall prove that if $U_6 \neq U_6^*$ then one of the equations (37) has a solution ξ_B in K_6. Let $\varepsilon \in U_6 \setminus U_6^*$. Since $<\!-1\!>N_{6/3}(U_6^*) = U_3$ there exists an $\omega \in U_6^*$ such that $N_{6/3}(\omega) = \pm N_{6/3}(\varepsilon)$. Dividing by ω and by a relevant power of μ we may assume that

(38) $$N_{6/3}(\varepsilon) = \pm1, \quad N_{6/2}(\varepsilon) = \pm\mu^e$$

where $e \in \{-1,0,1\}$. The case $e = 0$ does not occur, because $\varepsilon \notin U_R \subset U_6^*$. Since $\varepsilon^3/\mu^e \in U_R$, we have

(39) $$\varepsilon^3 = \pm\mu^e\xi_R^h\xi_R'^k.$$

Dividing by relevant powers of ξ_R and ξ_R' we may assume that $h,k \in \{0,1,2\}$. Applying the automorphism σ to (39) and using (25) we have

(40) $$\varepsilon'^3 = \pm\mu^{-e}\xi_R'^h\xi_R''^k = \pm\mu^{-e}\xi_R^{-k}\xi_R'^{h+k}.$$

Hence from (39) and (40)

$$(41) \qquad (\varepsilon\varepsilon')^3 = \pm \xi_R^{h-k} \xi_R'^{h+2k}.$$

It follows that $h \equiv k \bmod 3$ so that $h = k$. If $h = k = 0$ then $\varepsilon^3 = \pm\mu^e$ which is impossible by Theorem 2. If $h = k = 1$ then the equation $x^3 = \mu^e \xi_R \xi_R'$ has the solution $\xi_B = \pm\varepsilon$. If $h = k = 2$ then the equation $x^3 = \mu^{-e} \xi_R \xi_R'$ has the solution $\xi_B = \pm\xi_R \xi_R'/\varepsilon$. This proves the assertion above. The theorem now follows because the last statement in (ii) is a consequence of Theorem 4. $\square$

$\underline{\text{Theorem 7.}}$ Let $<-1>N_{6/3}(U_6^*) \neq U_3$ and $N_{6/2}(U_6^*) = U_2$.

(i)   If one of the equations

$$(42) \qquad x^2 = |\tau\xi_R|, \quad x^2 = |\tau'\xi_R|, \quad x^2 = |\tau\tau'\xi_R|$$

has a solution $x = \xi_C \in K_6$, then $[U_6 : U_6^*] = 4$,
$U_6 = U_6^* \cup \xi_C U_6^* \cup \xi_C' U_6^* \cup \xi_C \xi_C' U_6^*$ and $<-1>N_{6/3}(U_6) = U_3$.

(ii) If none of the equations (42) has a solution in K_6, then $U_6 = U_6^*$ and $[U_3 : <-1>N_{6/3}(U_6)] = 4$.

$\underline{\text{Remark.}}$ Only one of the equations (42) can have a solution in K_6. Suppose, namely, that $\omega_1^2 = |\tau^c\tau'^d\xi_R|$ and $\omega_2^2 = |\tau^e\tau'^f\xi_R|$ where c, d, e, $f \in \{0,1\}$. Then $(\omega_1/\omega_2)^2 = |\tau^{c-e}\tau'^{d-f}|$. Now we cannot have $(c,d) \neq (e,f)$, according to Theorem 2.

$\underline{\text{Proof.}}$ Suppose first that one of the equations (42) has a solution $\xi_C \in K_6$. We have $N_{6/3}(\xi_C) = \pm\tau, \pm\tau'$, or $\pm\tau\tau'$, and $N_{6/2}(\xi_C) = \pm 1$. Again one can easily verify, using Theorem 4 and the argument in the proof of Theorem 5, that (i) holds.

We shall secondly show that, if $U_6 \neq U_6^*$ then one of the equations (42) has a solution ξ_C in K_6. Let $\varepsilon \in U_6 \setminus U_6^*$. Since $N_{6/2}(U_6^*) = U_2$ there exists an $\omega \in U_6^*$ such that $N_{6/2}(\omega) = N_{6/2}(\varepsilon)$. Dividing by ω and by relevant powers of τ and τ' we may assume that

$$N_{6/3}(\varepsilon) = \pm\tau^c\tau'^d, \quad N_{6/2}(\varepsilon) = 1$$

where $c,d \in \{0,1\}$. The case $c = d = 0$ does not occur, because $\varepsilon \notin U_R$ $\subset U_6^*$. Since $\varepsilon^2/(\tau^c\tau'^d) \in U_R$ we have

$$\varepsilon^2 = \pm\tau^c\tau'^d\xi_R^h\xi_R'^k.$$

Dividing by relevant powers of ξ_R and ξ_R' we may assume that $h,k \in \{0,1\}$. We cannot have $h = k = 0$ by Theorem 2. For $h = 1$, $k = 0$ the equation $x^2 = |\tau^c\tau'^d\xi_R|$ has the solution $\xi_C = \varepsilon$. For $h = 0$, $k = 1$ the equation $x^2 = |\tau^{2-c-d}\tau'^c\xi_R|$ has the solution $\xi_C = \varepsilon^v\tau^{1-d}\tau'^c$. For $h = k = 1$ the equation $x^2 = |\tau^d\tau'^{2-c-d}\xi_R|$ has the solution $\xi_C = \varepsilon^{iv}\xi_R'\tau^d\tau'^{1-c}$. This proves the assertion and the theorem follows as before. $\square$

$\underline{\text{Theorem 8.}}$ Let $<-1>N_{6/3}(U_6^*) \neq U_3$ and $N_{6/2}(U_6^*) \neq U_2$.

(i) If one of the equations (37) has a solution $\xi_B \in K_6$ and one of the equations (42) has a solution $\xi_C \in K_6$, then $[U_6 : U_6^*] = 12$,
$$U_6 = \bigcup_{\substack{i=0,1,2 \\ j=0,1 \\ k=0,1}} \xi_B^i\xi_C^j\xi_C'^kU_6^*, \quad <-1>N_{6/3}(U_6) = U_3 \text{ and } N_{6/2}(U_6) = U_2.$$

(ii) If one of the equations (37) has a solution $\xi_B \in K_6$ and none of the equations (42) has a solution in K_6, then $[U_6 : U_6^*] = 3$,
$$U_6 = U_6^* \cup \xi_BU_6^* \cup \xi_B^2U_6^*, \quad [U_3 : <-1>N_{6/3}(U_6)] = 4 \text{ and } N_{6/2}(U_6) = U_2.$$

(iii) If neither of the equations (37) has a solution in K_6 and one of the equations (42) has a solution $\xi_C \in K_6$, then $[U_6 : U_6^*] = 4$,
$$U_6 = U_6^* \cup \xi_CU_6^* \cup \xi_C'U_6^* \cup \xi_C\xi_C'U_6^*, \quad <-1>N_{6/3}(U_6) = U_3 \text{ and }$$
$[U_2 : N_{6/2}(U_6)] = 3$.

(iv) If none of the equations (37) and (42) has a solution in K_6, then
$$U_6 = U_6^*, \quad [U_3 : <-1>N_{6/3}(U_6)] = 4 \text{ and } [U_2 : N_{6/2}(U_6)] = 3.$$

$\underline{\text{Proof.}}$ By looking at the relative norms one can verify as in the proofs of Theorems 6 and 7 that the cosets U_6^*, $\xi_BU_6^*$, $\xi_B^2U_6^*$ are mutually disjoint if ξ_B exists, and so are the cosets U_6^*, $\xi_CU_6^*$, $\xi_C'U_6^*$, $\xi_C\xi_C'U_6^*$ if ξ_C exists. Furthermore the same argument shows that the cosets $\xi_B^i\xi_C^j\xi_C'^kU_6^*$ ($i \in \{0,1,2\}$; $j,k \in \{0,1\}$) are mutually disjoint if ξ_B and ξ_C both exist.

Let us suppose that $U_6 \neq U_6^*$ and take $\varepsilon \in U_6 \smallsetminus U_6^*$. Again we may assume that $N_{6/3}(\varepsilon) = \pm \tau^c \tau'^d$ where $c,d \in \{0,1\}$, and $N_{6/2}(\varepsilon) = \pm \mu^e$ where $e \in \{-1,0,1\}$. If $c = d = 0$, we can prove the existence of ξ_B in the same way as in the proof of Theorem 6. Depending on the value of e we have $\varepsilon U_6^* = \xi_B U_6^*$ or $\xi_B^2 U_6^*$. Similarly, if $e = 0$ then ξ_C exists and $\varepsilon U_6^* = \xi_C U_6^*$, $\xi_C' U_6^*$, or $\xi_C \xi_C' U_6^*$. Let us therefore suppose that $e \neq 0$ and at least one of the numbers c,d is $\neq 0$. Since $\varepsilon^6/(\mu^{2e} \tau^{3c} \tau'^{3d}) \in U_R$ we can write

(43)
$$\varepsilon^6 = \pm \mu^{2e} \tau^{3c} \tau'^{3d} \xi_R^h \xi_R'^k$$

from which it further follows that

(44)
$$(\varepsilon^2/(\mu^e \tau^c \tau'^d))^3 = \pm \mu^{-e} \xi_R^h \xi_R'^k.$$

As in the proof of Theorem 6 we deduce that ξ_B exists. Clearly $\varepsilon^2 U_6^* = \xi_B^1 U_6^*$ ($1 = 1$ or 2). From (43) we also obtain

$$(\varepsilon^3/(\mu^e \tau^c \tau'^d))^2 = \pm \tau^c \tau'^d \xi_R^h \xi_R'^k.$$

Arguing as in the proof of Theorem 7 we find that ξ_C exists. Further $\varepsilon^3 U_6^* = \xi_C^i \xi_C'^j U_6^*$ ($i,j \in \{0,1\}$; $i + j > 0$). Hence $\varepsilon U_6^* = \xi_B^{-1} \xi_C^i \xi_C'^j U_6^*$ $= \xi_B^{3-1} \xi_C^i \xi_C'^j U_6^*$. The theorem now follows immediately from these considerations. □

It is evident that if $U_6 = U_6^*$ and $\xi_R \in \langle -1, \mu, \tau, \tau', \xi_A, \xi_A' \rangle$ then $\{\mu, \tau, \tau', \xi_A, \xi_A'\}$ is a system of fundamental units of K_6. From Latimer [17] we know that K_6 has a system of fundamental units containing μ, τ, τ'. We shall give a direct proof for this result.

Theorem 9. The field K_6 has a system of fundamental units of the form $\{\mu, \tau, \tau', \varepsilon_1, \varepsilon_2\}$.

Proof. It is well known that we may choose fundamental units ω_1, ω_2, ω_3, ω_4, ω_5 of K_6 so that

(45)
$$\tau = \pm\omega_1^{a_{11}},$$

(46)
$$\tau' = \pm\omega_1^{a_{21}}\omega_2^{a_{22}},$$

(47)
$$\mu = \pm\omega_1^{a_{31}}\omega_2^{a_{32}}\omega_3^{a_{33}},$$

where the a_{ij} are integers and $a_{ii} > 0$, $0 \leqslant a_{ij} < a_{ii}$ ($1 \leqslant j < i \leqslant 3$). From Theorem 2 we obtain $a_{11} = 1$. So we may choose $\omega_1 = \tau$. Now $\pm\tau'\tau^{-a_{21}} = \omega_2^{a_{22}}$. Again from Theorem 2 we have that $a_{22} = 1$. So we may choose $\omega_2 = \tau'$. Taking norms in (47) we have

(48)
$$\mu^3 = \pm(N_{6/2}(\omega_3))^{a_{33}},$$

(49)
$$\pm 1 = \tau^{2a_{31}}\tau'^{2a_{32}}(N_{6/3}(\omega_3))^{a_{33}}.$$

From (48) we have $a_{33} = 1$ or 3. If $a_{33} = 3$, then from (49) we obtain $3 | a_{31}$ and $3 | a_{32}$. But now $\mu = \pm(\tau^{a_{31}/3}\tau'^{a_{32}/3}\omega_3)^3$ which is impossible according to Theorem 2. So $a_{33} = 1$ and we may take $\omega_3 = \mu$. □

Let $\mathbb{Q}(\beta)/\mathbb{Q}$ be an Abelian extension of degree n. Let the conjugates of β be $\beta_1, \ldots, \beta_n$. We define

(50)
$$\mathcal{M}(\beta) = \frac{1}{n} \sum_{i=1}^{n} |\beta_i|^2 .$$

In every finite extension K of $\mathbb{Q}(\beta)$ the set of conjugates of β consists of $\beta_1, \ldots, \beta_n$ each one appearing $[K : \mathbb{Q}(\beta)]$ times. So also in the field K, $\mathcal{M}(\beta)$ equals the mean value of the squares of the absolute values of the conjugates of β. The conjugates of the number $|\beta|^2$ in $\mathbb{Q}(\beta)$ are $|\beta_1|^2, \ldots, |\beta_n|^2$, for the Galois group of the extension $\mathbb{Q}(\beta)/\mathbb{Q}$ is commutative. Hence $\mathcal{M}(\beta) \in \mathbb{Q}$. As indicated by Loxton [21, p. 166]

(51)
$$\mathcal{M}(\beta^k) \geqslant \mathcal{M}(\beta)^k \quad (k \in \mathbb{Z}, \ k \geqslant 0).$$

If β is an algebraic integer then $n\mathcal{M}(\beta)$ is the trace of the algebraic integer $|\beta|^2$ from $\mathbb{Q}(\beta)$ to $\mathbb{Q}$. So $n\mathcal{M}(\beta)$ is an integer.

We shall now study the function $\mathcal{M}$ in the group U_R. Let $\varepsilon \in U_R$. Then there exist $\alpha \in K_3$ and $\beta \in K_3$ such that $\varepsilon = \alpha + \beta\sqrt{m}$. From (22) we have

(52)
$$N_{6/3}(\varepsilon) = (\alpha + \beta\sqrt{m})(\alpha - \beta\sqrt{m}) = \alpha^2 - m\beta^2 = 1.$$

Now

$$
\begin{aligned}
6\mathcal{M}(\varepsilon) = {} & (\alpha + \beta\sqrt{m})^2 + (\alpha' - \beta'\sqrt{m})^2 + (\alpha'' + \beta''\sqrt{m})^2 \\
& + (\alpha - \beta\sqrt{m})^2 + (\alpha' + \beta'\sqrt{m})^2 + (\alpha'' - \beta''\sqrt{m})^2 \\
= {} & 2S_{3/1}(\alpha^2) + 2mS_{3/1}(\beta^2).
\end{aligned}
$$

From (52) it follows that $S_{3/1}(\alpha^2) = mS_{3/1}(\beta^2) + 3$. Hence

(53)
$$\mathcal{M}(\varepsilon) = \frac{2m}{3} S_{3/1}(\beta^2) + 1.$$

Put $x = \varepsilon^2$ and $y = \varepsilon''^2$. Then from (23) and (25) we have

(54)
$$6\mathcal{M}(\varepsilon) = x + xy + y + \frac{1}{x} + \frac{1}{xy} + \frac{1}{y}.$$

In the next theorem we get a lower bound for the function $\mathcal{M}$ in $U_R \smallsetminus \{\pm 1\}$. We use the notation

(55)
$$f_* = (f_2, f_3), \quad f_2' = \frac{f_2}{f_*}, \quad f_3' = \frac{f_3}{f_*}.$$

Theorem 10. If $\varepsilon \in U_R \smallsetminus \{\pm 1\}$ then

(56)
$$\mathcal{M}(\varepsilon) \geqslant \begin{cases} \dfrac{f_2(2f_3 + f_*)}{18f_*} + 1 & \text{when } 3 \nmid f_* \\[4mm] \dfrac{f_2(2f_3 + 3f_*)}{54f_*} + 1 & \text{when } 3 \mid f_*. \end{cases}$$

Proof. Let $\varepsilon \in U_R \smallsetminus \{\pm 1\}$ and let $\varepsilon = \alpha + \beta\sqrt{m}$ where $\alpha, \beta \in K_3$. Replacing ε by $-\varepsilon$, if necessary, we may assume that $N_{6/2}(\varepsilon) = 1$. According to Theorem 1, $\beta = \frac{1}{f_*}(z_0 + z_1\theta + z_2\theta')$ if $2 \mid f_2$ and $\beta = \frac{1}{2f_*}(z_0 + z_1\theta + z_2\theta')$ if $2 \nmid f_2$, where the z_i are integers. Put $\gamma = z_0 + z_1\theta + z_2\theta'$ in both cases. From (53) it now follows that

$$\mathcal{M}(\varepsilon) = \frac{f_2}{6f_*^2} S_{3/1}(\gamma^2) + 1.$$

Using (15) it further follows that

(57) $$\mathcal{M}(\varepsilon) = \frac{f_2}{18f_*^2}(S_{3/1}(\gamma))^2 + \frac{f_2 f_3}{9f_*^2}(z_1^2 + z_2^2 - z_1 z_2) + 1.$$

If $z_1 = z_2 = 0$ then $\beta \in \mathbb{Q}$. Now (52) implies that α is of degree 2 or less over the field $\mathbb{Q}$. Since $\alpha \in K_3$, we would have $\alpha \in \mathbb{Q}$ and $\alpha + \beta\sqrt{m} \in K_2$. So $\varepsilon^3 = N_{6/2}(\varepsilon) = 1$, a contradiction. Hence $z_1^2 + z_2^2$

$- z_1 z_2 > 0.$

Since $\beta \sqrt{m} = \frac{1}{2f_*} \gamma \sqrt{f_2}$, we have

(58)
$$\gamma = \frac{\sqrt{f_2}}{f_2'} (\varepsilon - \varepsilon''')$$

from which we obtain

(59)
$$S_{3/1}(\gamma) = \frac{\sqrt{f_2}}{f_2'} (\varepsilon - \varepsilon''' - \varepsilon' + \varepsilon^{iv} + \varepsilon'' - \varepsilon^v)$$

$$= \frac{\sqrt{f_2}}{f_2'} N_{6/2}(\varepsilon - 1).$$

Since $\varepsilon \neq 1$, we have $S_{3/1}(\gamma) \neq 0$. Since $S_{3/1}(\gamma) \in \mathbb{Z}$ and $N_{6/2}(\varepsilon - 1) \in \mathcal{O}_2$, $N_{6/2}(\varepsilon - 1) = x\sqrt{m}$ where x is an integer. So $f_* | S_{3/1}(\gamma)$. Hence we get from (57)

(60)
$$\mathcal{M}(\varepsilon) \geq \frac{f_2}{18} + \frac{f_2 f_3}{9 f_*^2} (z_1^2 + z_2^2 - z_1 z_2) + 1.$$

Let p be a prime divisor of f_*. Then p is ramified both in the extension $K_2/\mathbb{Q}$ and in the extension $K_3/\mathbb{Q}$. Hence p is fully ramified in the extension $K_6/\mathbb{Q}$. So $p = \mathfrak{p}^6$ where $\mathfrak{p}$ is a prime ideal of the ring $\mathcal{O}_6$. Since $\mathfrak{p}^6 | S_{3/1}(\gamma)$ and $\mathfrak{p}^3$ is the highest power of $\mathfrak{p}$ which divides $\sqrt{f_2}$, (59) implies that ε has a conjugate $\varepsilon^{(i)}$ such that $\varepsilon^{(i)} \equiv 1 \bmod \mathfrak{p}$. Since $\sigma^j(\mathfrak{p}) = \mathfrak{p}$ for all $j \in \mathbb{Z}$, we have

$$\varepsilon^{(i)} \equiv 1 \bmod \mathfrak{p} \quad (i = 0, \ldots, 5).$$

From (58) we obtain

$$\gamma^2 = \frac{f_*}{f_2'} (\varepsilon - \varepsilon''')^2.$$

Since $\varepsilon - \varepsilon''' \equiv 0 \bmod \mathfrak{p}$, $p|f_*$ and $p \nmid f_2'$, $p^2 | S_{3/1}(\gamma^2)$. Since p was any prime divisor of f_*, $f_*^2 | S_{3/1}(\gamma^2)$. Now (57) implies $z_1^2 + z_2^2 - z_1 z_2 \equiv 0 \bmod f_*/(3, f_*)$. Hence we obtain from (60)

$$\mathcal{M}(\varepsilon) \geq \frac{f_2}{18} + \frac{f_2 f_3}{9 f_* (3, f_*)} + 1$$

which is the same as (56). ☐

Next we shall prove two theorems concerning monotonicity of the function $\mathcal{M}$. Before it, however, we prove another useful theorem.

Theorem 11. If $\varepsilon \in U_R \smallsetminus \{\pm 1\}$, then there exists exactly one integer t such that

(61) $$1 < |\varepsilon^{(t)}| < |\varepsilon^{(t+1)}| \ , \quad 0 \leqslant t \leqslant 5.$$

Proof. Let an integer t be such that $|\varepsilon^{(t+1)}| = \max_{0 \leqslant i \leqslant 5} |\varepsilon^{(i)}|$, $0 \leqslant t \leqslant 5$. According to (25), $|\varepsilon^{(t)}| = |\varepsilon^{(t+1)}|$ implies that $|\varepsilon^{(t+2)}| = 1$. This is impossible, because $\varepsilon \neq \pm 1$. So

$$|\varepsilon^{(t)}| < |\varepsilon^{(t+1)}|.$$

Since $|\varepsilon^{(t)}||\varepsilon^{(t+2)}| = |\varepsilon^{(t+1)}| > |\varepsilon^{(t+2)}|$,

$$1 < |\varepsilon^{(t)}|.$$

Thus t satisfies the condition (61).

Conversely, let us suppose that an integer t satisfies (61). From (25) we obtain

$$1 < |\varepsilon^{(t+2)}| < |\varepsilon^{(t+1)}|.$$

From (23) it further follows that

$$|\varepsilon^{(t+i)}| < 1 \quad (i = 3,4,5).$$

Hence $|\varepsilon^{(t+1)}| = \max_{0 \leqslant i \leqslant 5} |\varepsilon^{(i)}|$. ☐

Theorem 12. If $\varepsilon \in U_R \smallsetminus \{\pm 1\}$ then

(62) $$\mathcal{M}(\varepsilon^k \varepsilon'^{l+1}) > \mathcal{M}(\varepsilon^k \varepsilon'^l)$$

for all integers k, l such that $k \geqslant 0$, $l \geqslant 0$.

Proof. Put $a = \varepsilon^2$ and $b = \varepsilon''^2$. According to Theorem 11, ε can be chosen among it's conjugates so that $1 < |\varepsilon| < |\varepsilon'|$. So $a > 1$ and from

(25) we also have b > 1. According to (54) and (25)

(63) $6\mathcal{M}(\varepsilon^k\varepsilon'^{1+1}) = a^{k+1+1}b^{1+1} + a^k b^{k+1+1} + a^{-1-1}b^k$

$+ a^{-k-1-1}b^{-1-1} + a^{-k}b^{-k-1-1} + a^{1+1}b^{-k}$

and

(64) $6\mathcal{M}(\varepsilon^k\varepsilon'^1) = a^{k+1}b^1 + a^k b^{k+1} + a^{-1}b^k$

$+ a^{-k-1}b^{-1} + a^{-k}b^{-k-1} + a^1 b^{-k}.$

Put

$$\Delta = 6a^{k+1+1}b^{k+1+1}\{\mathcal{M}(\varepsilon^k\varepsilon'^{1+1}) - \mathcal{M}(\varepsilon^k\varepsilon'^1)\}.$$

The statement (62) will be proved if we verify that $\Delta > 0$. From the equations (63) and (64) we obtain

(65) $\Delta = (a^{k+21+1}b^{1+1} - a^k b^{2k+1+1})(a - 1)$

$+ (a^{2k+1+1}b^{2k+21+1} - a^{1+1})(b - 1)$

$+ (a^{2k+21+1}b^{k+21+1} - b^k)(ab-1).$

Let Δ_1, Δ_2, Δ_3 denote the terms of the sum (65) in order. Now $\Delta_2 > 0$ and $\Delta_3 > 0$, for a > 1, b > 1, k $\geq$ 0, 1 $\geq$ 0. If a $\leq$ b, then $\Delta_2 > -\Delta_1$. If a > b, then $\Delta_3 > -\Delta_1$. Hence $\Delta > 0$. □

Theorem 13. If $\varepsilon \in U_R \smallsetminus \{\pm 1\}$, then

(66) $$\mathcal{M}(\varepsilon^{k+1}\varepsilon'^1) > \mathcal{M}(\varepsilon^k\varepsilon'^1)$$

for all integers k, 1 such that k $\geq$ 0, 1 $\geq$ 0.

Proof. We use the same method as in the proof of Theorem 12. So let $a = \varepsilon^2$, $b = \varepsilon''^2$ and we may assume that a > 1, b > 1. Furthermore

$6a^{k+1+1}b^{k+1+1}\{\mathcal{M}(\varepsilon^{k+1}\varepsilon'^1) - \mathcal{M}(\varepsilon^k\varepsilon'^1)\} =$

$(b^{1+2k+1}a^{k+1} - b^1 a^{21+k+1})(b - 1) + (b^{21+k+1}a^{21+2k+1} - b^{k+1})(a - 1)$

$+ (b^{21+2k+1}a^{1+2k+1} - a^1)(ab - 1).$

Interchanging k and 1, and a and b this formula is the same as Δ in (65). Thus, according to the proof of Theorem 12, (66) is valid. □

Since $6\mathcal{M}(\varepsilon)$ is a positive integer for all $\varepsilon \in U_R$, there exists a least number in the set $\{\mathcal{M}(\varepsilon) \mid \varepsilon \in U_R \smallsetminus \{\pm 1\}\}$. Let $\mathcal{M}(\xi_R)$ be such a number. The next theorem shows that ξ_R is a generating relative unit. In Leopoldt [19] a more general result is proved in a different way.

<u>Theorem 14.</u> Let $\xi_R \in U_R \smallsetminus \{\pm 1\}$ be such that $\mathcal{M}(\xi_R) =$ min $\{\mathcal{M}(\varepsilon) \mid \varepsilon \in U_R \smallsetminus \{\pm 1\}\}$. Then

$$U_R = \{\pm \xi_R{}^k \xi_R'{}^l \mid k, l \in \mathbb{Z}\}.$$

Furthermore, every $\varepsilon \in U_R$ has exactly one expression in the form $\varepsilon = (-1)^\nu \xi_R{}^k \xi_R'{}^l$ where $\nu \in \{0, 1\}$ and k, l are integers.

<u>Proof.</u> According to Theorem 11 we may assume that $1 < |\xi_R| < |\xi_R'|$. From (25) we also have $1 < |\xi_R''| < |\xi_R'|$. Put $a = \xi_R{}^2$ and $b = \xi_R''{}^2$. So $a > 1$ and $b > 1$. Let ε be any element in U_R. The determinant of the pair of equations

(67)
$$\begin{cases} k \ln|\xi_R| + l \ln|\xi_R'| = \ln|\varepsilon| \\ k \ln|\xi_R'| + l \ln|\xi_R''| = \ln|\varepsilon'| \end{cases}$$

is $\ln|\xi_R| \ln|\xi_R''| - \ln|\xi_R'| \ln|\xi_R'| \neq 0$. So (67) has a unique real solution k, l. Thus there exist unique real numbers k, l such that $\varepsilon = \pm \xi_R{}^k \xi_R'{}^l$ and $\varepsilon' = \pm \xi_R'{}^k \xi_R''{}^l$. Now the theorem will be proved if we show that k and l are integers. Let $k = K + x$ and $l = L + y$ where K, $L \in \mathbb{Z}$ and $0 \leqslant x, y < 1$. Put $\varepsilon_1 = \xi_R{}^x \xi_R'{}^y$. Since $\varepsilon_1 = \pm \varepsilon \xi_R{}^{-K} \xi_R'{}^{-L}$, $\varepsilon_1 \in U_R$. Now, according to (54),

(68)
$$\begin{aligned} 6\mathcal{M}(\varepsilon_1) = {}& a^{x+y} b^y + a^x b^{x+y} + a^{-y} b^x \\ & + a^{-x-y} b^{-y} + a^{-x} b^{-x-y} + a^y b^{-x}. \end{aligned}$$

Let $f(x, y)$ denote the right side of the equation (68). Let us examine the function f in the region $0 \leqslant x, y \leqslant 1$, $x + y \leqslant 1$. Clearly $f_{xx}(x, y) > 0$ in this region. So f has, when y is fixed, an absolute maximum in the interval $0 \leqslant x \leqslant 1 - y$ at $x = 0$ or at $x = 1 - y$ and not at other points. Now

$$f(0,y) = (ab)^y + b^y + a^{-y} + (ab)^{-y} + b^{-y} + a^y$$

and

$$f(1-y,y) = ab^y + a^{1-y}b + a^{-y}b^{1-y} + a^{-1}b^{-y} + a^{-1+y}b^{-1} + a^yb^{-1+y}.$$

Since the function $h(x) = x + 1/x$ is strictly increasing in the interval $x \geqslant 1$, $f(0,y)$ has an absolute maximum in the interval $0 \leqslant y \leqslant 1$ at exactly one point $y = 1$. Clearly $\dfrac{d^2f(1-y,y)}{dy^2} > 0$. So $f(1-y,y)$ has an absolute maximum in the interval $0 \leqslant y \leqslant 1$ at $y = 0$ or at $y = 1$ and not at other points. According to (68), $f(0,1) = f(1,0) = 6\mathcal{M}(\xi_R)$. Thus $f(x,y) < 6\mathcal{M}(\xi_R)$ when $0 \leqslant x,y < 1$, $x + y \leqslant 1$. Since $\varepsilon_1 \in U_R$, then because of the choice of ξ_R, $x = y = 0$ or $x + y > 1$. Let us suppose that $x + y > 1$. Then $6\mathcal{M}(\xi_R\xi_R'/\varepsilon_1) = f(1-x,1-y)$ and $0 < 1-x$, $1-y < 1$, $(1-x) + (1-y) < 1$. Now, according to the preceding proof, $6\mathcal{M}(\xi_R\xi_R'/\varepsilon_1) < 6\mathcal{M}(\xi_R)$, a contradiction. Hence $x = y = 0$ and so k and l are integers. □

In this work a computer program is constructed for finding a generating relative unit ξ_R. First a candidate ξ_0 for ξ_R is chosen in the following way:

$$(69) \qquad \xi_0 = \begin{cases} \mu^{-w/3}\, \tau^{-u/2}\, \tau'^{-v/2}\xi_A & \text{if } 3|w,\ 2|u,\ 2|v \\[6pt] \mu^{-2w/3}\, \tau^{-u}\, \tau'^{-v}\, \xi_A^2 & \text{if } 3|w, \text{ and } 2\!\nmid\!u \text{ or } 2\!\nmid\!v \\[6pt] \tau^{(v-u)/2}\, \tau'^{-u/2}\, \xi_A\xi_A' & \text{if } 3\!\nmid\!w,\ 2|u,\ 2|v \\[6pt] \tau^{v-u}\, \tau'^{-u}\, \xi_A^2\xi_A'^2 & \text{if } 3\!\nmid\!w, \text{ and } 2\!\nmid\!u \text{ or } 2\!\nmid\!v \end{cases}$$

where u, v, w are the integers in (34) and (35). In Sections 6 and 7 we shall derive from Bergström's product formula [13, p. 59] equations for computing the co-ordinates of ξ_A. If $2\!\nmid\!u$ or $2\!\nmid\!v$ then we have from Theorem 2 that $\pm\xi_0$ is not a square in K_6. If $2|u$ and $2|v$, then in many cases Theorem 3 and the co-ordinates of ξ_0 may be used to determine if $\pm\xi_0$ is not a square in K_6. If $\pm\xi_0$ is a square then we replace it by its square root and so on until we reach a relative unit ξ_1 such that $\xi_1^{2^n} = \pm\xi_0$ for some $n \geqslant 0$ and $\pm\xi_1$ is not a square in K_6. (For $n = 0$, $\xi_1 = \xi_0$.)

Whether ξ_1 is a generating relative unit or not will be found out in the following way. Let us assume that $\xi_1 = \pm\xi_R^K \xi_R'^L$ for some integers K, L. We may suppose that $K > 0$, $L \geqslant 0$. If for instance $K = 0$, then we replace ξ_R by its conjugate ξ_R', and if $K < 0$ then we replace ξ_R by ξ_R''. So we have $K > 0$. If $L < 0$, let $J = \min\{K, -L\}$. From (23) and (24) we have now

$$\xi_1 = \pm\xi_R^K \xi_R''^{iv\ -L}\xi_R''^J \xi_R^{vJ} = \pm\xi_R^{K-J} \xi_R''^{iv\ -L-J}\xi_R^{vJ}.$$

If $J = K \neq -L$ then we replace ξ_R by ξ_R^{iv}, and if $J = -L$ then we replace ξ_R by ξ_R^v. Hence we also have $L \geqslant 0$. From Theorem 10 we get a lower bound M_{min} for $\mathcal{M}(\xi_R)$, viz.

$$(70) \qquad M_{min} = \begin{cases} \dfrac{f_2(2f_3 + f_*)}{18f_*} + 1 & \text{if } 3 \nmid f_* \\[2em] \dfrac{f_2(2f_3 + 3f_*)}{54f_*} + 1 & \text{if } 3 \mid f_* \end{cases}.$$

According to Theorems 12 and 13, and (51), we have

$$(71) \qquad \mathcal{M}(\xi_1) = \mathcal{M}(\xi_R^K \xi_R'^L) \geqslant \mathcal{M}(\xi_R^K) \geqslant \mathcal{M}(\xi_R)^K \geqslant M_{min}^K,$$

$$(72) \qquad \mathcal{M}(\xi_1) = \mathcal{M}(\xi_R^K \xi_R'^L) \geqslant \mathcal{M}(\xi_R'^L) \geqslant \mathcal{M}(\xi_R')^L \geqslant M_{min}^L.$$

Put

$$(73) \qquad K_{max} = [\ln\mathcal{M}(\xi_1)/\ln M_{min}]$$

where [ ] denotes the greatest integer function. From (71) and (72) we obtain K, L $\leqslant K_{max}$. Since $\pm\xi_1$ is not a square in K_6, both K and L cannot be even. Now

$$\xi_1^L \xi_1'^K = \pm\xi_R^{KL}\xi_R'^{K^2+L^2}\xi_R''^{KL} = \pm\xi_R'^{K^2+KL+L^2}$$

and, since $K^2 + KL + L^2$ is odd,

$$(74) \qquad \pm\xi_R' = \xi_1^{L/(K^2+KL+L^2)}\ \xi_1'^{K/(K^2+KL+L^2)}.$$

Now we have

(75) $\qquad S_{6/1}(\pm\xi_R^!) = \sum_{i=0}^{5} \sigma^i(\xi_1)^{L/(K^2+KL+L^2)} \; \sigma^i(\xi_1^!)^{K/(K^2+KL+L^2)}.$

Because $\xi_R^! \in \mathcal{O}_6$, $S_{6/1}(\pm\xi_R^!)$ is an integer. Thus if the right side of (75) is not an integer for all integers K, L such that $0 \leqslant K,L \leqslant K_{max}$, $K + L > 1$ and $K^2 + KL + L^2$ odd, then ξ_1 already is a generating relative unit of K_6. Otherwise the right side of (74) is such a unit for suitable values of K and L provided that it belongs to $\mathcal{O}_6$.

5. Bergström's product formula

Let K be a real Abelian number field of degree n and let f be the conductor of K. Let H be the multiplicative group of prime residue classes mod f corresponding to K, according to class field theory. If $x + f\mathbb{Z} \in H$ then also $-x + f\mathbb{Z} \in H$, because K is real. Choose one residue class from every pair $x + f\mathbb{Z}$, $-x + f\mathbb{Z}$ and then take an odd representative from each such residue class. Let α be the set of these representatives and let l be the number of elements in α. Then $l = \varphi(f)/2n$. The number

$$\xi = \prod_{x \in \alpha}(\zeta_{2f}^{x} - \zeta_{2f}^{-x})$$

is an integer of $\mathbb{Q}(\zeta_{2f})$. Let $(r + f\mathbb{Z})H$, where r is odd, be any element of the factor group Φ_f/H. Put

$$\xi_r = \prod_{x \in \alpha}(\zeta_{2f}^{rx} - \zeta_{2f}^{-rx}).$$

The numbers ξ/ξ_r belong to the field K [14, p. 22] and they are called the cyclotomic units of K.

Hasse has proved Bergström's product formula in his work [13]. In Bergström's product formula the number ξ is represented by means of Gaussian sums. The needed representation of ξ as a sum of roots of unity we obtain in the following way. Let $\alpha = \{a_1, \ldots, a_l\}$. Then

$$\xi = \zeta_{2f}^{\sum_{i=1}^{l} a_i} \prod_{i=1}^{l}(1 - \zeta_{2f}^{-2a_i}).$$

Multiplying this product term by term we obtain the sum

$$\xi = \sum_{t=1}^{2f} B_t \zeta_{2f}^{\ t},$$

where the coefficients $B_t = B_t^{(1)}$ are obtained from the recursion formula

$$(76) \qquad \begin{cases} B_t^{(0)} = \begin{cases} 1 \text{ when } t \equiv \sum\limits_{i=1}^{1} a_i \mod 2f \\ 0 \text{ otherwise} \end{cases} \\ B_t^{(i)} = B_t^{(i-1)} - B_{t+2a_i}^{(i-1)} \quad (i = 1, \ldots, 1). \end{cases}$$

In (76) the indexes t of $B_t^{(i)}$ have to be taken modulo $2f$. Since the numbers a_i are odd, $B_t = 0$ if $t \not\equiv 1 \mod 2$. For an odd f we have

$$(77) \qquad \xi = \sum_{t=1}^{f} A_t \zeta_f^{\ t} \text{ where } A_t = \begin{cases} B_{2t} \text{ if } 1 \text{ is even} \\ -B_{2t+f} \text{ if } 1 \text{ is odd}. \end{cases}$$

If f and 1 are even then we have

$$(78) \qquad \xi = \sum_{t=1}^{f} A_t \zeta_f^{\ t} \text{ where } A_t = \frac{1}{2}(B_{2t} - B_{2t+f}).$$

If f is even and 1 is odd then

$$(79) \qquad \xi = \sum_{\substack{t=1 \\ t \text{ odd}}}^{2f-1} A_t \zeta_{2f}^{\ t} \text{ where } A_t = \frac{1}{2}(B_t - B_{t+f}).$$

We shall now consider the character which Hasse [13, p. 56] denotes by ψ. Put

$$\bar{H} = \{ x + 2f\mathbb{Z} \mid x + f\mathbb{Z} \in H, x \equiv 1 \mod 2 \}.$$

If f is odd, then $\#\bar{H} = \#H$ and furthermore $\bar{H}$ is isomorphic to H. If f is even, then $\#\bar{H} = 2 \cdot \#H$. For every $x + 2f\mathbb{Z} \in \bar{H}$ the automorphism $\zeta_{2f} \mapsto \zeta_{2f}^{\ x}$ of the field $\mathbb{Q}(\zeta_{2f})$ maps ξ into ξ or into $-\xi$. The condition

$$\xi \mapsto \psi(x)\xi$$

defines a character ψ of the group $\overline{H}$. The character ψ is either principal or quadratic. According to Hasse [13, p. 56] every x such that
$x + 2f\mathbb{Z} \in \overline{H}$ satisfies a congruence

$$x^1 \equiv (-1)^\nu + \delta f \bmod 2f$$

where $\nu = 0$ or 1 and $\delta = 0$ or 1. Furthermore ψ is determined for every
x such that $x + 2f\mathbb{Z} \in \overline{H}$ as follows:

$$(80) \qquad\qquad \psi(x) = \begin{cases} 1 \text{ when } x^1 \equiv 1, \; -1-f \bmod 2f \\ -1 \text{ when } x^1 \equiv -1, \; 1+f \bmod 2f. \end{cases}$$

According to Hasse [13, p. 57], $\xi \in K$ if and only if ψ is principal.
Otherwise only $\xi^2 \in K$. Let us now examine ψ in different cases.

Case 1. f odd, 1 odd. According to (80), $\psi(-1) = -1$. Thus ψ is
a quadratic character. The character group of $\overline{H}$ is isomorphic to $\overline{H}$.
Since $\#\overline{H} = 21$ and 1 is odd, there exists a unique element of order 2
in $\overline{H}$. Thus in the character group of $\overline{H}$ there is a unique quadratic
character and ψ must be that character. We shall consider ψ as a character of H via the canonical isomorphism $\overline{H} \approx H$.

Case 2. f odd, 1 even. a) Suppose first that the 2-Sylow subgroup
of $\overline{H}$ is not cyclic. Then the order of any element in $\overline{H}$ divides $\#\overline{H}/2 =$
1. So $x^1 \equiv 1 \bmod 2f$ for all x such that $x + 2f\mathbb{Z} \in \overline{H}$. By (80), ψ is
the principal character.

b) Suppose next that the 2-Sylow subgroup of $\overline{H}$ is cyclic. Let $x +$
$2f\mathbb{Z}$ be a generator of that cyclic group. We cannot have $x^1 \equiv \pm(1+f)$
mod 2f because f is odd. Since the order of $x + 2f\mathbb{Z}$ in $\overline{H}$ does not
divide 1 we must have $x^1 \equiv -1 \bmod 2f$. Hence ψ is quadratic. In this
case $\overline{H}$ also has a unique quadratic character which must be ψ.

As in Case 1 we shall consider ψ as a character of H.

Case 3. f even, 1 odd. According to (80), $\psi(-1) = -1$ so that ψ
is a quadratic character. Since

$$(-1-f)^1 = (-1)^1(1+f)^1 \equiv -1-f \bmod 2f,$$

we have $\psi(-1-f) = 1$. So ψ can be regarded as a character of the factor

group $\overline{H}/<-1-f + 2f\mathbb{Z} > = \overline{H}_{-1-f}$, say. Since $\#\overline{H}_{-1-f} = \#\overline{H}/2 = 21$ and 1 is

odd, ψ is the unique quadratic character of $\overline{H}_{-1-f}$. Here we have $\psi(1+f)$

$= -1$ so that ψ cannot be considered as a character of H.

Case 4. f even, 1 even. We have $\psi(-1) = 1$ so that ψ can be re-

garded as a character of the factor group $\overline{H}/<-1 + 2f\mathbb{Z} > = \overline{H}_{-1}$, say.

a) If the 2-Sylow subgroup of $\overline{H}_{-1}$ is not cyclic the order of any

element of $\overline{H}_{-1}$ divides 1. Hence $x^1 \equiv \pm 1 \bmod 2f$ for all x such that

$x + 2f\mathbb{Z} \in \overline{H}$. We cannot have $x^1 \equiv -1 \bmod 2f$ because f and 1 are even.

So ψ is the principal character.

b) Suppose now that the 2-Sylow subgroup of $\overline{H}_{-1}$ is cyclic and let

$(x + 2f\mathbb{Z})<-1 + 2f\mathbb{Z} >$ be a generator of that cyclic group. The order

of this element in $\overline{H}_{-1}$ does not divide 1. Therefore $x^1 \not\equiv \pm 1 \bmod 2f$.

Since $-1-f \equiv 3 \bmod 4$, $x^1 \equiv -1-f \bmod 2f$ is impossible. So $x^1 \equiv 1+f \bmod$

2f whence ψ is quadratic. The group $\overline{H}_{-1}$ has a unique quadratic charac-

ter which must be ψ.

From (80), $\psi(1+f) = 1$ whence we can (and shall) consider ψ as a

character of H.

For every positive divisor d of f let

$$H^{(d)} = \{ x + \tfrac{f}{d}\mathbb{Z} \mid x + f\mathbb{Z} \in H \},$$

m(d) be the index of $H^{(d)}$ in $\Phi_{f/d}$ and

$$\Phi_{f/d}/H^{(d)} = \{ s_1^{(d)}H^{(d)}, s_2^{(d)}H^{(d)}, \ldots, s_{m(d)}^{(d)}H^{(d)} \}.$$

Correspondingly for every positive divisor d of 2f let

$$\overline{H}^{(d)} = \{ x + \tfrac{2f}{d}\mathbb{Z} \mid x + 2f\mathbb{Z} \in \overline{H} \},$$

$\overline{m}(d)$ be the index of $\overline{H}^{(d)}$ in $\Phi_{2f/d}$ and

$$\overline{\Phi}_{2f/d}/\overline{H}^{(d)} = \{ s_1^{(d)}\overline{H}^{(d)}, s_2^{(d)}\overline{H}^{(d)}, \ldots , s_{\overline{m}(d)}^{(d)}\overline{H}^{(d)}\}.$$

We shall now give Bergström's product formula in different cases. We use the notation $f(\chi)$ for the conductor of a character χ, $\tau(\chi)$ for the Gaussian sum belonging to χ and μ for the Möbius function.

Case 1. f odd, or f and 1 even.   In this case

$$(81) \quad \xi = \frac{1}{n} \sum_{\chi} \sum_{d|\frac{f}{f(\chi)}} \frac{n}{\overline{m}(d)} \mu(\frac{f}{f(\chi)d}) \chi(\frac{f}{f(\chi)d}) \sum_{x=1}^{m(d)} \overline{\chi}(s_x^{(d)}) A_{s_x^{(d)}d} \tau(\chi)$$

where the summation $\sum_{\chi}$ is over all extensions of ψ to the group $\overline{\Phi}_f$ and the numbers A_t are obtained from (77) or (78).

Case 2. f even and 1 odd.   In this case

$$(82) \quad \xi = \frac{1}{n} \sum_{\chi} \sum_{d|\frac{2f}{f(\chi)}} \frac{n}{\overline{m}(d)} \mu(\frac{2f}{f(\chi)d}) \chi(\frac{2f}{f(\chi)d}) \sum_{x=1}^{\overline{m}(d)} \overline{\chi}(s_x^{(d)}) A_{s_x^{(d)}d} \tau(\chi)$$

where the summation $\sum_{\chi}$ is over all extensions of ψ to the group $\overline{\Phi}_{2f}$ and the numbers A_t are obtained from (79).

6. Bergström's product formula in the
case of real cyclic sextic fields

According to Hasse [13, p. 60], in the case of real cyclic field K_n a sufficient condition for ψ to be the principal character is that the conductor f_n of K_n is decomposable. We say that f_n is decomposable if it can be decomposed into a product of two nontrivial relatively prime factors $f_n = f'f''$ so that in the decomposition of a generating character $\chi_n = \chi'\chi''$, where the conductors of χ' and χ'' are f' and f'' respectively, the characters χ' and χ'' are even.

We shall now examine, when the conductor f_6 of K_6 is decomposable. A generating character of K_6 is $\chi_6 = \chi_2\chi_3$. We recall from p. 6 that $f_3 = p_0 p_1 \cdots p_n$ or $9p_1 \cdots p_n$ where p_i is a prime $\equiv 1 \bmod 3$ ($i = 0,1,\ldots,n$) and $p_i \neq p_j$ ($i \neq j$). The number ϕ defined on p. 6 decomposes into a product

(83)
$$\phi = \begin{cases} \pi_0' \pi_1' \cdots \pi_n' & \text{when } 3 \nmid f_3 \\ 3\rho^\alpha \pi_1' \cdots \pi_n' & \text{when } 3 \mid f_3 \end{cases}$$

where $\pi_i' = (a_i \pm b_i\sqrt{-3})/2$, $(a_i^2 + 3b_i^2)/4 = p_i$, $a_i \equiv 2 \bmod 3$, $b_i > 0$, $b_i \equiv 0 \bmod 3$ and $\alpha = \pm 1$. The character χ_3' defined by the equation

(84)
$$\chi_3' = \begin{cases} \left(\dfrac{\cdot}{\pi_0'}\right)_3 \left(\dfrac{\cdot}{\pi_1'}\right)_3 \cdots \left(\dfrac{\cdot}{\pi_n'}\right)_3 & \text{when } 3 \nmid f_3 \\ \left(\dfrac{\cdot}{\rho}\right)_3^\alpha \left(\dfrac{\cdot}{\pi_1'}\right)_3 \cdots \left(\dfrac{\cdot}{\pi_n'}\right)_3 & \text{when } 3 \mid f_3 \end{cases}$$

is either χ_3 or $\bar{\chi}_3$ by the notation on p. 9 . The values of the cubic residue symbol $\left(\dfrac{\cdot}{\pi_i'}\right)_3$ are obtained from the congruence

$$(85) \qquad \left(\frac{x}{\pi_i'}\right)_3 \equiv x^{(p_i-1)/3} \quad \bmod \pi_i'$$

and the values of the cubic residue symbol $\left(\frac{\rho}{x}\right)_3$ are obtained from the equation

$$(86) \qquad \left(\frac{\rho}{x}\right)_3 = \rho^{(x^2-1)/3}.$$

If a character χ_n decomposes into the product of two or more characters such that the conductors of these are relatively prime, we shall use for these characters the notation $\chi_{n,f}$ where f denotes the conductor of the character. So $\left(\frac{x}{\pi_i'}\right)_3 = \chi_{3,p_i}$ or $\bar{\chi}_{3,p_i}$ and $\left(\frac{\rho}{x}\right)_3 = \chi_{3,9}$ or $\bar{\chi}_{3,9}$. From (85) and (86) we see that the characters χ_{3,p_i} (i = 0,1, ... ,n) and $\chi_{3,9}$ are even. Thus, if $f_3 \nmid 3f_2$ we have $\chi_6 = (\chi_2 \chi_{3,k})\chi_{3,f_3/k}$ where f_3/k and the conductor of $\chi_2 \chi_{3,k}$ are relatively prime and $f_3/k \neq 1$. Hence f_6 is decomposable if $f_3 \nmid 3f_2$. If $f_3 \mid 3f_2$, then f_6 is decomposable if and only if f_2 is decomposable.

The conductor f_2 of K_2 is of the form

$$(87) \qquad f_2 = \begin{cases} q_0 q_1 \cdots q_k & \text{if } f_2 \equiv 1 \bmod 4 \\ 4q_1 \cdots q_k, \; q_1 \cdots q_k \equiv 3 \bmod 4 & \text{if } f_2 \equiv 4 \bmod 8 \\ 8q_1 \cdots q_k & \text{otherwise} \end{cases}$$

where the q_i are distinct odd primes. Let $\chi_{2,4}'$ and $\chi_{2,8}'$ be the characters defined for odd values of x as follows

$$(88) \qquad \chi_{2,4}'(x) = (-1)^{(x-1)/2},$$

$$(89) \qquad \chi_{2,8}'(x) = (-1)^{(x^2-1)/8}.$$

So the character χ_2 is

$$(90) \quad X_2 = \begin{cases} X_{2,q_0} X_{2,q_1} \cdots X_{2,q_t} & \text{when } f_2 = q_0 q_1 \cdots q_t \equiv 1 \bmod 4 \\[2ex] X'_{2,4} X_{2,q_1} \cdots X_{2,q_t} & \text{when } f_2 = 4 q_1 \cdots q_t \text{ and} \\ & \qquad q_1 \cdots q_t \equiv 3 \bmod 4 \\[2ex] X'_{2,8} X_{2,q_1} \cdots X_{2,q_t} & \text{when } f_2 = 8 q_1 \cdots q_t \text{ and} \\ & \qquad q_1 \cdots q_t \equiv 1 \bmod 4 \\[2ex] X'_{2,8} X'_{2,4} X_{2,q_1} \cdots X_{2,q_t} & \text{when } f_2 = 8 q_1 \cdots q_t \text{ and} \\ & \qquad q_1 \cdots q_t \equiv 3 \bmod 4 \end{cases}$$

where X_{2,q_i} is the Legendre symbol $\left(\dfrac{\cdot}{q_i}\right)$ $(i = 0, 1, \ldots, t)$. The character X_{2,q_i} is even if $q_i \equiv 1 \bmod 4$, and odd if $q_i \equiv 3 \bmod 4$. The character $X'_{2,4}$ is odd and $X'_{2,8}$ is even. Hence f_2 is decomposable except in the following cases:

$$(91) \quad \begin{cases} 1) \ f_2 = q_0 \text{ where } q_0 \equiv 1 \bmod 4, \\[1ex] 2) \ f_2 = q_0 q_1 \text{ where } q_0, \ q_1 \equiv 3 \bmod 4, \\[1ex] 3) \ f_2 = 4 q_1 \text{ where } q_1 \equiv 3 \bmod 4, \\[1ex] 4) \ f_2 = 8, \\[1ex] 5) \ f_2 = 8 q_1 \text{ where } q_1 \equiv 3 \bmod 4. \end{cases}$$

According to the preceding results, the cases in which f_6 is not decomposable, i.e. the cases in which $f_3 | 3 f_2$ and f_2 is not decomposable, are (by 1) we denote the decomposable case)

2) $f_6 = f_2 = f_3 = p_0$ where $p_0 \equiv 1 \bmod 12$,

3) $f_6 = f_2 = p_0 p_1$, $f_3 = p_0$ where $p_0 \equiv 7 \bmod 12$, $p_1 \equiv 3 \bmod 4$,

4) $f_6 = 9 p_1$, $f_2 = 3 p_1$, $f_3 = 9$ where $p_1 \equiv 3 \bmod 4$,

5) $f_6 = f_2 = f_3 = p_0 p_1$ where $p_0, \ p_1 \equiv 7 \bmod 12$,

6) $f_6 = 9 p_1$, $f_2 = 3 p_1$, $f_3 = 9 p_1$ where $p_1 \equiv 7 \bmod 12$,

7) $f_6 = f_2 = 4 p_0$, $f_3 = p_0$ where $p_0 \equiv 7 \bmod 12$,

8) $f_6 = 4 \cdot 9$, $f_2 = 4 \cdot 3$, $f_3 = 9$,

9) $f_6 = f_2 = 8 p_0$, $f_3 = p_0$ where $p_0 \equiv 7 \bmod 12$,

10) $f_6 = 8 \cdot 9$, $f_2 = 8 \cdot 3$, $f_3 = 9$.

The group H is

$$H = \{ x + f_6 \mathbb{Z} \mid \chi_3(x) = 1, \chi_2(x) = 1 \}$$

[14, p. 5]. From p. 35 we recall that for any positive divisor d of f_6

$$H^{(d)} = \{ x + \frac{f_6}{d}\mathbb{Z} \mid x + f_6\mathbb{Z} \in H \}$$

and m(d) is the index of $H^{(d)}$ in $\Phi_{f_6/d}$. From p. 33 we recall that

$$\bar{H} = \{ x + 2f_6\mathbb{Z} \mid x + f_6\mathbb{Z} \in H, x \equiv 1 \bmod 2 \}.$$

For any positive divisor d of $2f_6$

$$\bar{H}^{(d)} = \{ x + \frac{2f_6}{d}\mathbb{Z} \mid x + 2f_6\mathbb{Z} \in \bar{H} \}$$

and $\bar{m}(d)$ is the index of $\bar{H}^{(d)}$ in $\Phi_{2f_6/d}$. We can take s, defined on p. 12, to be any odd integer such that $\chi_2(s) = -1$ and $\chi_3(s) = \rho$.

<u>Theorem 15.</u> Let d be any positive divisor of f_6. Then the group $\Phi_{f_6/d}/H^{(d)}$ is generated by $(s + (f_6/d)\mathbb{Z})H^{(d)}$, and

$$m(d) = \begin{cases} 6 & \text{if } d = 1 \\ 3 & \text{if } d \mid f_2', d > 1 \\ 2 & \text{if } d \mid f_3', d > 1 \\ 1 & \text{if } d \nmid f_2', d \nmid f_3' . \end{cases}$$

<u>Proof.</u> If $x + (f_6/d)\mathbb{Z} \in \Phi_{f_6/d}$ then $(x + jf_6/d, f_6) = 1$ for some j. Furthermore $x + jf_6/d + f_6\mathbb{Z} \in (s + f_6\mathbb{Z})^l H$ for some l. So $x + (f_6/d)\mathbb{Z} \in (s + (f_6/d)\mathbb{Z})^l H^{(d)}$, which proves the first statement. Hence m(d) is the least positive integer k such that $s^k + (f_6/d)\mathbb{Z} \in H^{(d)}$.

In the case d = 1 the assertion is trivial. Suppose that $d \mid f_2'$, d > 1. Since the conductor f_2 of χ_2 does not divide f_6/d there is an integer g such that $g \equiv 1 \bmod f_6/d$, $\chi_2(g) = -1$. So $\chi_2(s^3 g) = 1$. Further $\chi_3(s^3 g) = 1$ because f_3 does divide f_6/d. Therefore m(d) = 1 or 3. But $\chi_3(t) = \rho$ for every t satisfying $t \equiv s \bmod f_6/d$, whence m(d) = 3.

In the case $d \mid f_3'$, $d > 1$ the argument is similar. Suppose now that $d \nmid f_2'$, $d \nmid f_3'$. In this case neither f_2 nor f_3 divides f_6/d. There exist g and h such that $(gh, f_6) = 1$, $g \equiv 1 \bmod f_6/d$, $h \equiv 1 \bmod f_6/d$, $\chi_2(g) = -1$, $\chi_3(h) = \rho^{-e}$ ($e = 1$ or 2). We have $\chi_2(sg^3h^{4e}) = \chi_3(sg^3h^{4e}) = 1$. Hence $m(d) = 1$. $\square$

Theorem 16. Let d be any positive divisor of $2f_6$. Then the group $\Phi_{2f_6/d}/\overline{H}^{(d)}$ is generated by $(s + (2f_6/d)\mathbb{Z})\overline{H}^{(d)}$, and

$$
\overline{m}(d) = \begin{cases}
6 & \text{if } d \leqslant 2 \\
3 & \text{if } d \mid 2f_2', \ d > 2 \\
2 & \text{if } d \mid 2f_3', \ d > 2 \\
1 & \text{if } d \nmid 2f_2', \ d \nmid 2f_3' \ .
\end{cases}
$$

Proof. Case 1. d odd. We shall now show that $\Phi_{2f_6/d}/\overline{H}^{(d)}$ is isomorphic to $\Phi_{f_6/d}/H^{(d)}$. The function $x + (2f_6/d)\mathbb{Z} \mapsto (x + (f_6/d)\mathbb{Z})H^{(d)}$ from $\Phi_{2f_6/d}$ into $\Phi_{f_6/d}/H^{(d)}$ is surjective, because we can choose an odd representative for $x + (f_6/d)\mathbb{Z} \in \Phi_{f_6/d}$. The kernel of the function is $\overline{H}^{(d)}$, because the following statements are equivalent for $(x, 2f_6) = 1$

(i) $x + (2f_6/d)\mathbb{Z}$ belongs to the kernel of the function,

(ii) $x + (f_6/d)\mathbb{Z} \in H^{(d)}$,

(iii) $x \equiv x' \bmod f_6/d$ for some x' such that $x' + f_6\mathbb{Z} \in H$,

(iv) $x \equiv x'' \bmod 2f_6/d$ for some x'' such that $x'' + f_6\mathbb{Z} \in H$,

(v) $x \equiv x'' \bmod 2f_6/d$ for some x'' such that $x'' + 2f_6\mathbb{Z} \in \overline{H}$.

The result follows now from Theorem 15.

Case 2. d even. Using the equivalence of (iv) and (v) we have $\overline{H}^{(d)} = H^{(d/2)}$. Hence $\Phi_{2f_6/d}/\overline{H}^{(d)} = \Phi_{2f_6/d}/H^{(d/2)}$ and the result follows from Theorem 15. $\square$

Next we shall determine the needed extensions of ψ and give Bergström's product formula in different cases. We shall need the following properties of Gaussian sums. If the conductors f' and f" of the characters χ' and χ'' respectively are relatively prime, then

(92) $$\tau(\chi'\chi'') = \chi'(f'')\chi''(f')\tau(\chi')\tau(\chi'')$$

[15, p. 452]. If p is an odd prime $\equiv 1 \mod 3$ and $\chi_{3,p} = \left(\frac{}{\pi}\right)_3$ where $\pi \in \mathbb{Z}[\rho]$, $\pi\bar{\pi} = p$, $\pi \equiv 1 \mod 3$, then by [15, pp. 469, 480]

(93) $$\tau(\chi_{2,p}\chi_{3,p}) = -\frac{1}{p} \bar{\pi} \chi_{3,p}(2)\tau(\chi_{2,p})\tau(\chi_{3,p}).$$

Note that the number π in [15] satisfies $\pi \equiv 2 \mod 3$ which causes the minus sign in (93). If $\chi_3 = \chi_3'$, defined in (84), then let $\pi_i = \pi_i'$, and if $\chi_3 = \bar{\chi}_3'$ then let $\pi_i = \bar{\pi}_i'$ (i = 0, 1, ... ,n).

Case 1. f_6 is decomposable. We recall from p. 37 that in this case ψ is the principal character. Hence the extensions of ψ to the group Φ_{f_6} are χ_1, χ_2, χ_3, $\bar{\chi}_3$, $\chi_6 = \chi_2\chi_3$, $\bar{\chi}_6 = \chi_2\bar{\chi}_3$. The representation of $\tau(\chi_2\chi_3)$ by means of $\tau(\chi_2) = \sqrt{f_2}$ and $\tau(\chi_3) = (-1)^n(\theta + \rho\theta' + \rho^2\theta'')$ is as follows.

a) $3 \nmid f_*$. Now $\chi_2\chi_3 = \chi_{2,f_2'} \chi_{2,f_*} \chi_{3,f_*} \chi_{3,f_3'}$ and $(f_2',f_*) = (f_2',f_3') = (f_*,f_3') = 1$. So, according to (92) and (93),

$$\tau(\chi_2\chi_3) = \chi_{2,f_2'}(f_3)\chi_{2,f_*}(f_2'f_3')\chi_{3,f_*}(f_2'f_3')\chi_{3,f_3'}(f_2) \cdot$$

$$\tau(\chi_{2,f_2'})\tau(\chi_{2,f_*}\chi_{3,f_*})\tau(\chi_{3,f_3'})$$

$$= \frac{1}{f_*} \prod_{p_i|f_*}(-\bar{\pi}_i) \chi_2(f_3')\chi_3(f_2')\chi_{3,f_*}(2)\tau(\chi_2)\tau(\chi_3).$$

b) $3 | f_*$. Now $\chi_2\chi_3 = \chi_{2,f_2'} \chi_{2,f_*} \chi_{3,3f_*} \chi_{3,f_3'/3}$ and $(f_2',3f_*) = (f_2',f_3'/3) = (3f_*,f_3'/3) = 1$. So

(94) $$\tau(\chi_2\chi_3) = \chi_{2,f_2'}(f_3)\chi_{2,f_*}(f_2'f_3'/3)\chi_{3,3f_*}(f_2'f_3'/3)\chi_{3,f_3'/3}(3f_2) \cdot$$

$$\tau(\chi_{2,f_2'})\tau(\chi_{2,f_*}\chi_{3,3f_*})\tau(\chi_{3,f_3'/3}).$$

Since $\tau(\chi_{2,3}\left(\frac{\rho}{}\right)_3) - \tau(\left(\frac{\rho}{}\right)_3) = -2(\rho\zeta_9^2 + \rho^2\zeta_9^5 + \zeta_9^8)$, $\zeta_9^3 = \rho$ and $\rho^2 + \rho + 1 = 0$, we have $\tau(\chi_{2,3}\left(\frac{\rho}{}\right)_3) = \tau(\left(\frac{\rho}{}\right)_3)$. Using the fact $\tau(\chi_{2,3}) = \sqrt{-3}$ we obtain

$$\tau(\chi_{2,3}\left(\frac{\rho}{}\right)_3) = -\frac{1}{3} \sqrt{-3} \tau(\chi_{2,3})\tau(\left(\frac{\rho}{}\right)_3).$$

So

(95)
$$\tau(\chi_{2,3}\chi_{3,9}) = (-1)^{\nu} \frac{1}{3} \sqrt{-3}\ \tau(\chi_{2,3})\tau(\chi_{3,9})$$

where $\nu = 1$ if $\chi_{3,9} = \left(\frac{\varrho}{}\right)_3$, and $\nu = 0$ if $\bar{\chi}_{3,9} = \left(\frac{\varrho}{}\right)_3$. From (92), (93), (94) and (95) we have

$$\tau(\chi_2\chi_3) = (-1)^{\nu} \frac{1}{f_*} \sqrt{-3} \prod_{P_i|f_*/3} (-\bar{\pi}_i)\ \chi_2(f_3'/3)\chi_{2,f_2/3}(3)\chi_{3}(f_2')\cdot$$

$$\chi_{3,f_*/3}(2)\tau(\chi_2)\tau(\chi_3).$$

Let $\tau(\chi_2\chi_3) = \frac{1}{f_*} T\sqrt{f_2}\,\tau(\chi_3)$ in both cases. Then $\tau(\chi_2\bar{\chi}_3) = \frac{1}{f_*} \bar{T}\sqrt{f_2}\,\tau(\bar{\chi}_3)$. Bergström's product formula (81) is now

(96) $6\xi =$
$$\mu(f_6)\sum_{x=0}^{5} A_{s^x} + 2\sum_{\substack{d|f_2'\\ d>1}}\mu(f_6/d)\sum_{x=0}^{2} A_{ds^x} + 3\sum_{\substack{d|f_3'\\ d>1}}\mu(f_6/d)\sum_{x=0}^{1} A_{ds^x}$$

$$+ 6\sum_{\substack{d|f_6\\ d\nmid f_2'\\ d\nmid f_3'}}\mu(f_6/d)A_d + \tau(\chi_3)\mu(f_2')\chi_3(f_2')\sum_{x=0}^{5}\rho^{2x}A_{s^x}$$

$$+ 2\tau(\chi_3)\sum_{\substack{d|f_2'\\ d>1}}\mu(f_2'/d)\chi_3(f_2'/d)\sum_{x=0}^{2}\rho^{2x}A_{ds^x}$$

$$+ \tau(\bar{\chi}_3)\mu(f_2')\bar{\chi}_3(f_2')\sum_{x=0}^{5}\rho^{x}A_{s^x}$$

$$+ 2\tau(\bar{\chi}_3)\sum_{\substack{d|f_2'\\ d>1}}\mu(f_2'/d)\bar{\chi}_3(f_2'/d)\sum_{x=0}^{2}\rho^{x}A_{ds^x}$$

$$+ \frac{1}{f_*}\sqrt{f_2}\left\{f_*\mu(f_3')\chi_2(f_3')\sum_{x=0}^{5}(-1)^{x}A_{s^x}\right.$$

$$+ 3f_*\sum_{\substack{d|f_3'\\ d>1}}\mu(f_3'/d)\chi_2(f_3'/d)\sum_{x=0}^{1}(-1)^{x}A_{ds^x}$$

$$\left.+ T\,\tau(\chi_3)\sum_{x=0}^{5}(-\rho^2)^{x}A_{s^x} + \bar{T}\,\tau(\bar{\chi}_3)\sum_{x=0}^{5}(-\rho)^{x}A_{s^x}\right\}.$$

Case 2. $f_6 = f_2 = f_3 = p_0$ $(p_0 \equiv 1 \mod 12)$. Since f_6 is odd, $\bar{H}$ is isomorphic to H. Let r be a primitive root modulo p_0. Then

$$H = \{\, r^x + p_0\mathbb{Z} \mid \chi_3(r^x) = 1,\ \chi_2(r^x) = 1 \,\}$$

$$= \{\, r^{6x} + p_0\mathbb{Z} \mid x = 0, 1, \ldots, 21-1 \,\} .$$

Thus H is cyclic. According to either Case 1 or Case 2b on p. 34, ψ is the quadratic character of H. Let χ_4 and $\bar{\chi}_4$ be the biquadratic characters, each of which has conductor p_0, and which are defined as follows

$$\chi_4(r) = i\ , \quad \bar{\chi}_4(r) = -i .$$

Since $\chi_4(r^6) = -1$ and $\bar{\chi}_4(r^6) = -1$, the characters χ_4 and $\bar{\chi}_4$ are extensions of ψ. Hence the extensions of ψ to the group Φ_{p_0} are χ_4, $\bar{\chi}_4$, $\chi_4\chi_3$, $\bar{\chi}_4\chi_3$, $\chi_4\bar{\chi}_3$, $\bar{\chi}_4\bar{\chi}_3$. Between the Gaussian sums $\tau(\chi_4'\chi_3'')$, $\tau(\chi_4')$, $\tau(\chi_3'')$, where $\chi_4' = \chi_4$ or $\bar{\chi}_4$ and $\chi_3'' = \chi_3$ or $\bar{\chi}_3$, there is a relation

$$\tau(\chi_4'\chi_3'') = \frac{1}{p_0}\,\bar{\pi}(\chi_4',\chi_3'')\,\tau(\chi_4')\,\tau(\chi_3''),$$

where $\pi(\chi_4',\chi_3'') = \sum\limits_{\substack{x,y\in\mathbb{Z} \\ 2\leqslant x\leqslant p_0-1 \\ x+y=1}} \chi_4'(x)\chi_3''(y)$ [15, pp. 462, 463]. From formula (81) we now get

$$(97)\ 6\xi = \frac{1}{p_0}\tau(\chi_4)\left\{ p_0\sum_{x=0}^{5}\bar{\chi}_4(s^x)A_{s^x} + \bar{\pi}(\chi_4,\chi_3)\tau(\chi_3)\sum_{x=0}^{5}\bar{\chi}_4(s^x)\rho^{2x}A_{s^x} \right.$$

$$\left. + \bar{\pi}(\chi_4,\bar{\chi}_3)\tau(\bar{\chi}_3)\sum_{x=0}^{5}\bar{\chi}_4(s^x)\rho^{x}A_{s^x} \right\}$$

$$+ \frac{1}{p_0}\tau(\bar{\chi}_4)\left\{ p_0\sum_{x=0}^{5}\chi_4(s^x)A_{s^x} + \bar{\pi}(\bar{\chi}_4,\chi_3)\tau(\chi_3)\sum_{x=0}^{5}\chi_4(s^x)\rho^{2x}A_{s^x} \right.$$

$$\left. + \bar{\pi}(\bar{\chi}_4,\bar{\chi}_3)\tau(\bar{\chi}_3)\sum_{x=0}^{5}\chi_4(s^x)\rho^{x}A_{s^x} \right\} .$$

Case 3. $f_6 = f_2 = p_0 p_1,\ f_3 = p_0$ $(p_0 \equiv 7 \mod 12,\ p_1 \equiv 3 \mod 4)$. Since f_6 and $1 = (p_0-1)(p_1-1)/12$ are odd, ψ is the quadratic character of H, according to Case 1 on p. 34. The characters χ_{2,p_0}, χ_{2,p_1}, $\chi_{2,p_0}\chi_3$,

$\chi_{2,P_0}\overline{\chi}_3$, $\chi_{2,P_1}\chi_3$, $\chi_{2,P_1}\overline{\chi}_3$ are odd. Hence they are the extensions of ψ to the group Φ_{f_6}. The Gaussian sums for these are

$$\tau(\chi_{2,P_j}) = i\sqrt{P_j} \quad (j = 0,1),$$

$$\tau(\chi_{2,P_0}\chi) = -i\frac{1}{P_0}\pi\,\chi(2)\sqrt{P_0}\,\tau(\chi) \quad ((\chi,\pi) = (\chi_3,\overline{\pi}_0), (\overline{\chi}_3,\pi_0)),$$

$$\tau(\chi_{2,P_1}\chi) = i\,\chi_{2,P_1}(P_0)\chi(P_1)\sqrt{P_1}\,\tau(\chi) \quad (\chi = \chi_3, \overline{\chi}_3).$$

Bergström's product formula (81) is now

$$(98) \quad 6\xi = i\sqrt{P_1}\left\{\tau(\chi_3)\chi_{2,P_1}(P_0)\chi_3(P_1)\sum_{x=0}^{5}\chi_{2,P_1}(s^x)\rho^{2x}A_{s^x}\right.$$

$$+ \tau(\overline{\chi}_3)\chi_{2,P_1}(P_0)\overline{\chi}_3(P_1)\sum_{x=0}^{5}\chi_{2,P_1}(s^x)\rho^{x}A_{s^x}$$

$$\left.- \chi_{2,P_1}(P_0)\sum_{x=0}^{5}\chi_{2,P_1}(s^x)A_{s^x} + 6A_{P_0}\right\}$$

$$+ i\frac{\sqrt{P_0}}{P_0}\left\{\tau(\chi_3)\,\overline{\pi}_0\,\chi_3(2P_1)\chi_{2,P_0}(P_1)\sum_{x=0}^{5}\chi_{2,P_0}(s^x)\rho^{2x}A_{s^x}\right.$$

$$- 2\tau(\chi_3)\,\overline{\pi}_0\,\chi_3(2)\sum_{x=0}^{2}\chi_{2,P_0}(s^x)\rho^{2x}A_{P_1 s^x}$$

$$+ \tau(\overline{\chi}_3)\,\pi_0\,\overline{\chi}_3(2P_1)\chi_{2,P_0}(P_1)\sum_{x=0}^{5}\chi_{2,P_0}(s^x)\rho^{x}A_{s^x}$$

$$- 2\tau(\overline{\chi}_3)\,\pi_0\,\overline{\chi}_3(2)\sum_{x=0}^{2}\chi_{2,P_0}(s^x)\rho^{x}A_{P_1 s^x}$$

$$\left.- P_0\,\chi_{2,P_0}(P_1)\sum_{x=0}^{5}\chi_{2,P_0}(s^x)A_{s^x} + 2P_0\sum_{x=0}^{2}\chi_{2,P_0}(s^x)A_{P_1 s^x}\right\}.$$

<u>Case 4. $f_6 = 9P_1$, $f_2 = 3P_1$, $f_3 = 9$ ($P_1 \equiv 3 \bmod 4$).</u> As in Case 3, $\chi_{2,3}$, χ_{2,P_1}, $\chi_{2,3}\chi_3$, $\chi_{2,3}\overline{\chi}_3$, $\chi_{2,P_1}\chi_3$, $\chi_{2,P_1}\overline{\chi}_3$ are the extensions of ψ to the group Φ_{f_6}. In this case

$$\tau(\chi_{2,3}) = i\sqrt{3}, \quad \tau(\chi_{2,P_1}) = i\sqrt{P_1}.$$

From (95) we have

$$\tau(\chi_{2,3}\chi) = (-1)^{\nu+1}\,\delta\,\tau(\chi) \quad ((\chi,\delta) = (\chi_3,1), (\overline{\chi}_3,-1))$$

where $\nu = 1$ if $\chi_3 = \left(\frac{\rho}{\cdot}\right)_3$ and $\nu = 0$ if $\overline{\chi}_3 = \left(\frac{\rho}{\cdot}\right)_3$. Furthermore

$$\tau(\chi_{2,p_1}\chi) = i\, \chi(p_1)\, \sqrt{p_1}\, \tau(\chi) \quad (\chi = \chi_3,\, \overline{\chi}_3).$$

From formula (81) we now obtain

$$(99)\quad 6\xi = i\, \sqrt{p_1}\, \left\{ \tau(\chi_3)\chi_3(p_1) \sum_{x=0}^{5} \chi_{2,p_1}(s^x)\rho^{2x} A_{s^x} \right.$$

$$+ \tau(\overline{\chi}_3)\overline{\chi}_3(p_1) \sum_{x=0}^{5} \chi_{2,p_1}(s^x)\rho^{x} A_{s^x}$$

$$\left. - 3\,\chi_{2,p_1}(3) \sum_{x=0}^{1} \chi_{2,p_1}(s^x) A_{3s^x} + 6A_9 \right\}$$

$$+ i\, \frac{1}{3}\, \sqrt{3}\, \left\{ (-1)^{\nu+1}\, \tau(\chi_3)\, \sqrt{-3}\; \chi_3(p_1)\chi_{2,3}(p_1) \sum_{x=0}^{5} \chi_{2,3}(s^x)\rho^{2x} A_{s^x} \right.$$

$$+ (-1)^{\nu}\, 2\tau(\chi_3)\, \sqrt{-3} \sum_{x=0}^{2} \chi_{2,3}(s^x)\rho^{2x} A_{p_1 s^x}$$

$$+ (-1)^{\nu}\, \tau(\overline{\chi}_3)\, \sqrt{-3}\; \overline{\chi}_3(p_1)\chi_{2,3}(p_1) \sum_{x=0}^{5} \chi_{2,3}(s^x)\rho^{x} A_{s^x}$$

$$+ (-1)^{\nu+1}\, 2\tau(\overline{\chi}_3)\, \sqrt{-3} \sum_{x=0}^{2} \chi_{2,3}(s^x)\rho^{x} A_{p_1 s^x}$$

$$\left. - 9\,\chi_{2,3}(p_1) \sum_{x=0}^{1} \chi_{2,3}(s^x) A_{3s^x} + 18A_{3p_1} \right\}.$$

<u>Case 5. $f_6 = f_2 = f_3 = p_0 p_1\ (p_0,\, p_1 \equiv 7 \bmod 12)$.</u>

The characters χ_{2,p_0}, χ_{2,p_1}, $\chi_{2,p_0}\chi_3$, $\chi_{2,p_0}\overline{\chi}_3$, $\chi_{2,p_1}\chi_3$, $\chi_{2,p_1}\overline{\chi}_3$ are the extensions of ψ to the group Φ_{f_6} as in Case 3. Now

$$\tau(\chi_{2,p_j}) = i\, \sqrt{p_j} \quad (j = 0,1),$$

$$\tau(\chi_{2,p_j}\chi_3) = -i\, \frac{1}{p_j}\, \overline{\pi}_j\, \chi_{3,p_j}(2)\, \chi_{2,p_j}(f_2/p_j)\, \sqrt{p_j}\, \tau(\chi_3) \quad (j = 0,1),$$

$$\tau(\chi_{2,p_j}\overline{\chi}_3) = -i\, \frac{1}{p_j}\, \pi_j\, \overline{\chi}_{3,p_j}(2)\, \chi_{2,p_j}(f_2/p_j)\, \sqrt{p_j}\, \tau(\overline{\chi}_3) \quad (j = 0,1).$$

Bergström's product formula (81) is now

$$(100) \quad 6\xi = i \frac{1}{p_0} \sqrt{p_0} \left\{ - \tau(\chi_3) \, \bar{\pi}_0 \, \chi_{2,p_0}(p_1) \, \chi_{3,p_0}(2) \sum_{x=0}^{5} \chi_{2,p_0}(s^x) \rho^{2x} A_{sx} \right.$$

$$- \tau(\bar{\chi}_3) \, \pi_0 \, \chi_{2,p_0}(p_1) \, \bar{\chi}_{3,p_0}(2) \sum_{x=0}^{5} \chi_{2,p_0}(s^x) \, \rho^x A_{sx}$$

$$\left. - p_0 \, \chi_{2,p_0}(p_1) \sum_{x=0}^{5} \chi_{2,p_0}(s^x) A_{sx} + 6p_0 A_{p_1} \right\}$$

$$+ i \frac{1}{p_1} \sqrt{p_1} \left\{ - \tau(\chi_3) \, \bar{\pi}_1 \, \chi_{2,p_1}(p_0) \, \chi_{3,p_1}(2) \sum_{x=0}^{5} \chi_{2,p_1}(s^x) \rho^{2x} A_{sx} \right.$$

$$- \tau(\bar{\chi}_3) \, \pi_1 \, \chi_{2,p_1}(p_0) \, \bar{\chi}_{3,p_1}(2) \sum_{x=0}^{5} \chi_{2,p_1}(s^x) \, \rho^x A_{sx}$$

$$\left. - p_1 \, \chi_{2,p_1}(p_0) \sum_{x=0}^{5} \chi_{2,p_1}(s^x) A_{sx} + 6p_1 A_{p_0} \right\} .$$

<u>Case 6.</u> $\underline{f_6 = 9p_1, f_2 = 3p_1, f_3 = 9p_1}$ $\underline{(p_1 \equiv 7 \bmod 12)}$.
Also in this case $\chi_{2,3}$, χ_{2,p_1}, $\chi_{2,3}\chi_3$, $\chi_{2,3}\bar{\chi}_3$, $\chi_{2,p_1}\chi_3$, $\chi_{2,p_1}\bar{\chi}_3$ are the extensions of ψ to the group Φ_{f_6}. The Gaussian sums for these are

$$\tau(\chi_{2,3}) = i \sqrt{3} , \quad \tau(\chi_{2,p_1}) = i \sqrt{p_1} ,$$

$$\tau(\chi_{2,3}\chi) = (-1)^{\nu+1} \delta \, \tau(\chi) \quad ((\chi,\delta) = (\chi_3, 1), (\bar{\chi}_3, -1)),$$

$$\tau(\chi_{2,p_1}\chi_3) = - i \frac{1}{p_1} \bar{\pi}_1 \, \chi_{3,p_1}(2) \sqrt{p_1} \, \tau(\chi_3),$$

$$\tau(\chi_{2,p_1}\bar{\chi}_3) = - i \frac{1}{p_1} \pi_1 \, \bar{\chi}_{3,p_1}(2) \sqrt{p_1} \, \tau(\bar{\chi}_3)$$

where ν is the number in (95). Now formula (81) becomes

$$(101) \quad 6\xi = i \frac{1}{3} \sqrt{3} \left\{ (-1)^{\nu} \tau(\chi_3) \sqrt{-3} \sum_{x=0}^{5} \chi_{2,3}(s^x) \rho^{2x} A_{sx} \right.$$

$$+ (-1)^{\nu+1} \tau(\bar{\chi}_3) \sqrt{-3} \sum_{x=0}^{5} \chi_{2,3}(s^x) \rho^x A_{sx}$$

$$\left. - 9 \sum_{x=0}^{1} \chi_{2,3}(s^x) A_{3sx} + 18 A_{3p_1} \right\}$$

$$+ i \frac{1}{p_1} \sqrt{p_1} \left\{ - \tau(\chi_3) \bar{\pi}_1 \chi_{3,p_1}(2) \sum_{x=0}^{5} \chi_{2,p_1}(s^x) \rho^{2x} A_{s^x} \right.$$

$$- \tau(\bar{\chi}_3) \pi_1 \bar{\chi}_{3,p_1}(2) \sum_{x=0}^{5} \chi_{2,p_1}(s^x) \rho^x A_{s^x}$$

$$\left. + 3p_1 \sum_{x=0}^{1} \chi_{2,p_1}(s^x) A_{3s^x} + 6p_1 A_9 \right\}.$$

<u>Case 7. $f_6 = f_2 = 4p_0, f_3 = p_0$ $(p_0 \equiv 7 \bmod 12)$.</u> Since f_6 is even and $1 = (p_0-1)/6$ is odd, ψ is the quadratic character of $\bar{H}_{-1-f_6}$, according to Case 3 on p. 34. The character $\chi'_{2,8}$ defined in (89) is even. The characters $\chi_{2,4}$ and χ_{2,p_0} are odd. Hence $\chi'_{2,8}\chi_{2,4}$, $\chi'_{2,8}\chi_{2,p_0}$, $\chi'_{2,8}\chi_{2,4}\chi_3$, $\chi'_{2,8}\chi_{2,4}\bar{\chi}_3$, $\chi'_{2,8}\chi_{2,p_0}\chi_3$, $\chi'_{2,8}\chi_{2,p_0}\bar{\chi}_3$ are quadratic characters of $\bar{H}$. Since $\chi'_{2,8}(-1-4p_0) = \chi'_{2,8}(3) = -1$, these characters are the extensions of ψ to the group Φ_{2f_6}. The Gaussian sums for these are

$$\tau(\chi'_{2,8}\chi_{2,4}) = i \, 2\sqrt{2}, \quad \tau(\chi'_{2,8}\chi_{2,p_0}) = i \, 2\sqrt{2p_0},$$

$$\tau(\chi'_{2,8}\chi_{2,4}\chi) = - i \, \chi'_{2,8}(p_0) \, 2\sqrt{2} \, \tau(\chi) \quad (\chi = \chi_3, \bar{\chi}_3),$$

$$\tau(\chi'_{2,8}\chi_{2,p_0}\chi) = - i \, \frac{1}{p_0} \pi \chi(2) \, 2\sqrt{2p_0} \, \tau(\chi) \quad ((\chi,\pi) = (\chi_3,\bar{\pi}_0), (\bar{\chi}_3,\pi_0)).$$

From Bergström's product formula (82) we now obtain

$$(102) \quad 6\xi = i \, 2\sqrt{2} \left\{ - \tau(\chi_3)\chi'_{2,8}(p_0) \sum_{x=0}^{5} \chi'_{2,8}(s^x)\chi_{2,4}(s^x)\rho^{2x} A_{s^x} \right.$$

$$- \tau(\bar{\chi}_3)\chi'_{2,8}(p_0) \sum_{x=0}^{5} \chi'_{2,8}(s^x)\chi_{2,4}(s^x)\rho^x A_{s^x}$$

$$\left. + \chi'_{2,8}(p_0) \sum_{x=0}^{5} \chi'_{2,8}(s^x)\chi_{2,4}(s^x)A_{s^x} + 6A_{p_0} \right\}$$

$$+ i \frac{2}{p_0} \sqrt{2p_0} \left\{ - \tau(\chi_3) \bar{\pi}_0 \chi_3(2) \sum_{x=0}^{5} \chi'_{2,8}(s^x)\chi_{2,p_0}(s^x) \rho^{2x} A_{s^x} \right.$$

$$- \tau(\bar{\chi}_3) \pi_0 \bar{\chi}_3(2) \sum_{x=0}^{5} \chi'_{2,8}(s^x)\chi_{2,p_0}(s^x) \rho^x A_{s^x}$$

$$+ p_0 \sum_{x=0}^{5} \chi'_{2,8}(s^x) \chi_{2,p_0}(s^x) A_{s^x} \Bigg\} .$$

Case 8. $f_6 = 4 \cdot 9$, $f_2 = 4 \cdot 3$, $f_3 = 9$. As in Case 7 the characters $\chi'_{2,8}\chi_{2,4}$, $\chi'_{2,8}\chi_{2,3}$, $\chi'_{2,8}\chi_{2,4}\chi_3$, $\chi'_{2,8}\chi_{2,4}\overline{\chi}_3$, $\chi'_{2,8}\chi_{2,3}\chi_3$, $\chi'_{2,8}\chi_{2,3}\overline{\chi}_3$ are the extensions of ψ to the group $\overline{\Phi}_{2f_6}$. Now

$$\tau(\chi'_{2,8}\chi_{2,4}) = i\, 2\sqrt{2}, \quad \tau(\chi'_{2,8}\chi_{2,3}) = i\, 2\sqrt{6},$$

$$\tau(\chi'_{2,8}\chi_{2,4}\chi) = i\, 2\sqrt{2}\,\tau(\chi) \quad (\chi = \chi_3,\ \overline{\chi}_3),$$

$$\tau(\chi'_{2,8}\chi_{2,3}\chi) = (-1)^\nu \delta\, 2\sqrt{2}\,\tau(\chi) \quad ((\chi,\delta) = (\chi_3, 1),\ (\overline{\chi}_3, -1))$$

where ν is the number in (95). Formula (82) is now

$$(103) \quad 6\xi = i\, 2\sqrt{2} \left\{ \tau(\chi_3) \sum_{x=0}^{5} \chi'_{2,8}(s^x) \chi_{2,4}(s^x) \rho^{2x} A_{s^x} \right.$$

$$+ \tau(\overline{\chi}_3) \sum_{x=0}^{5} \chi'_{2,8}(s^x) \chi_{2,4}(s^x) \rho^{x} A_{s^x}$$

$$\left. - 3\sum_{x=0}^{1} \chi'_{2,8}(s^x) \chi_{2,4}(s^x) A_{3s^x} + 6A_9 \right\}$$

$$+ i\, \frac{2}{3}\sqrt{6} \left\{ (-1)^{\nu+1} \tau(\chi_3)\, \sqrt{-3} \sum_{x=0}^{5} \chi'_{2,8}(s^x) \chi_{2,3}(s^x) \rho^{2x} A_{s^x} \right.$$

$$+ (-1)^\nu \tau(\overline{\chi}_3)\, \sqrt{-3} \sum_{x=0}^{5} \chi'_{2,8}(s^x) \chi_{2,3}(s^x) \rho^{x} A_{s^x}$$

$$\left. + 9\sum_{x=0}^{1} \chi'_{2,8}(s^x) \chi_{2,3}(s^x) A_{3s^x} \right\}$$

Case 9. $f_6 = f_2 = 8p_0$, $f_3 = p_0$ $(p_0 \equiv 7 \bmod 12)$. In this case f_6 and $1 = (p_0-1)/3$ are even. Now $\#\overline{H}_{-1} = 21 = 4(p_0-1)/6$ where $(p_0-1)/6$ is odd. The residue class $2p_0 - 1 + 16p_0\mathbb{Z}$ belongs to $\overline{H}$, for $\chi_3(2p_0-1) = \chi_3(-1) = 1$ and $\chi_2(2p_0-1) = \chi'_{2,8}(2p_0-1)\chi'_{2,4}(2p_0-1) \cdot$ $\chi_{2,p_0}(2p_0-1) = \chi'_{2,8}(5)\chi'_{2,4}(1)\chi_{2,p_0}(-1) = 1$. Since $(2p_0-1)^4 \equiv 8p_0(3p_0-1)$ $+ 1 \equiv 1 \bmod 16p_0$ and $(2p_0-1)^2 = 4p_0(p_0-1) + 1 \not\equiv \pm 1 \bmod 16p_0$, the 2-Sylow subgroup of $\overline{H}_{-1}$ is cyclic. According to Case 4b on p. 35 , ψ is the quadratic character of $\bar{H}_{-1}$. The characters $\chi'_{2,8}$, $\chi'_{2,4}\chi_{2,p_0}$, $\chi'_{2,8}\chi_3$, $\chi'_{2,8}\bar{\chi}_3$, $\chi'_{2,4}\chi_{2,p_0}\chi_3$, $\chi'_{2,4}\chi_{2,p_0}\bar{\chi}_3$ are quadratic characters of H. Since they are also even, they are the extensions of ψ to the group Φ_{f_6}. The Gaussian sums are

$$\tau(\chi'_{2,8}) = 2\sqrt{2}, \quad \tau(\chi'_{2,4}\chi_{2,p_0}) = 2\sqrt{p_0},$$

$$\tau(\chi'_{2,8}\chi) = \chi'_{2,8}(p_0) \, 2\sqrt{2} \, \tau(\chi) \quad (\chi = \chi_3, \bar{\chi}_3),$$

$$\tau(\chi'_{2,4}\chi_{2,p_0}\chi) = -\frac{1}{p_0} \pi \, 2\sqrt{p_0} \, \tau(\chi) \quad ((\chi,\pi) = (\chi_3, \bar{\pi}_0), (\bar{\chi}_3, \pi_0))$$

and Bergström's product formula (81) becomes

$$
(104) \quad 6\xi = 2\sqrt{2} \left\{ \tau(\chi_3)\chi'_{2,8}(p_0) \sum_{x=0}^{5} \chi'_{2,8}(s^x)\rho^{2x} A_{s^x} \right.
$$

$$
+ \tau(\bar{\chi}_3)\chi'_{2,8}(p_0) \sum_{x=0}^{5} \chi'_{2,8}(s^x)\rho^{x} A_{s^x}
$$

$$
\left. - \chi'_{2,8}(p_0) \sum_{x=0}^{5} \chi'_{2,8}(s^x) A_{s^x} + 6A_{p_0} \right\}
$$

$$
+ \frac{2}{p_0}\sqrt{p_0} \left\{ - 2\tau(\chi_3)\bar{\pi}_0 \sum_{x=0}^{2} \chi'_{2,4}(s^x)\chi_{2,p_0}(s^x)\,\rho^{2x} A_{2s^x} \right.
$$

$$
- 2\tau(\bar{\chi}_3)\pi_0 \sum_{x=0}^{2} \chi'_{2,4}(s^x)\chi_{2,p_0}(s^x)\,\rho^{x} A_{2s^x}
$$

$$
\left. + 2p_0 \sum_{x=0}^{2} \chi'_{2,4}(s^x)\chi_{2,p_0}(s^x)\, A_{2s^x} \right\}.
$$

<u>Case 10. $f_6 = 8\cdot 9$, $f_2 = 8\cdot 3$, $f_3 = 9$.</u> The group $\bar{H}_{-1}$ is cyclic, because the order of $(19 + 144\mathbb{Z})\langle -1 + 144\mathbb{Z}\rangle \in \bar{H}_{-1}$ is 4 and $\#\bar{H}_{-1} = 4$. According to Case 4b on p. 35, ψ is the quadratic character of $\bar{H}_{-1}$. The characters $\chi'_{2,8}$, $\chi'_{2,4}\chi_{2,3}$, $\chi'_{2,8}\chi_3$, $\chi'_{2,8}\bar{\chi}_3$, $\chi'_{2,4}\chi_{2,3}\chi_3$, $\chi'_{2,4}\chi_{2,3}\bar{\chi}_3$ are the extensions of ψ to the group Φ_{f_6} as in Case 9. The Gaussian sums of these are

$$\tau(\chi'_{2,8}) = 2\sqrt{2}, \quad \tau(\chi'_{2,4}\chi_{2,3}) = 2\sqrt{3},$$

$$\tau(\chi'_{2,8}\chi) = 2\sqrt{2}\,\tau(\chi) \quad (\chi = \chi_3, \bar{\chi}_3),$$

$$\tau(\chi'_{2,4}\chi_{2,3}\chi) = (-1)^{\nu+1} \delta \chi(4) \, 2i \, \tau(\chi) \quad ((\chi,\delta) = (\chi_3,1), (\overline{\chi}_3,-1))$$

where ν is the number in (95). Now from formula (81) we get

$$(105) \quad 6\xi = 2\sqrt{2} \left\{ \tau(\chi_3) \sum_{x=0}^{5} \chi'_{2,8}(s^x)\rho^{2x} A_{s^x} + \tau(\overline{\chi}_3) \sum_{x=0}^{5} \chi'_{2,8}(s^x) \rho^x A_{s^x} \right.$$

$$\left. + 3 \sum_{x=0}^{1} \chi'_{2,8}(s^x) A_{3s^x} + 6A_9 \right\}$$

$$+ \tfrac{2}{3}\sqrt{3} \left\{ (-1)^{\nu+1} 2\tau(\chi_3) \, \chi_3(4) \, \sqrt{-3} \sum_{x=0}^{2} \chi'_{2,4}(s^x)\chi_{2,3}(s^x) \rho^{2x} A_{2s^x} \right.$$

$$+ (-1)^{\nu} 2\tau(\overline{\chi}_3) \, \overline{\chi}_3(4) \, \sqrt{-3} \sum_{x=0}^{2} \chi'_{2,4}(s^x)\chi_{2,3}(s^x) \rho^x A_{2s^x}$$

$$\left. + 18A_6 \right\} .$$

Combining the definition of ξ_A on p. 16 and results on pp. 42 - 51 we get the following equivalent conditions:

(i)   Case 1 holds,

(ii) f_6 is decomposable,

(iii) ψ is principal,

(iv) $\xi \in K_6$,

(v) $\xi_A = \xi$,

(vi)   either $f_3 | 3f_2$ and f_2 does not have any of the forms (91) or $f_3 \nmid 3f_2$.

7. Formulas for computing ξ_A

Since $\tau(\chi_3) = (-1)^n(\theta + \rho\theta' + \rho^2\theta'')$ and $\tau(\bar{\chi}_3) = (-1)^n(\theta + \rho^2\theta' + \rho\theta'')$, we have

(106) $$\tau(\chi_3) + \tau(\bar{\chi}_3) = (-1)^n(-S_{3/1}(\theta) + 3\theta),$$

(107) $$\rho\tau(\chi_3) + \rho^2\tau(\bar{\chi}_3) = (-1)^n(2S_{3/1}(\theta) - 3\theta - 3\theta'),$$

(108) $$\rho^2\tau(\chi_3) + \rho\tau(\bar{\chi}_3) = (-1)^n(-S_{3/1}(\theta) + 3\theta')$$

where $S_{3/1}(\theta) = -1$ if $3 \nmid f_3$, and $S_{3/1}(\theta) = 0$ if $3 \mid f_3$. Using these equations and the equation

(109) $$\frac{c + d\sqrt{-3}}{2} = \frac{c + d}{2} + d\rho$$

we obtain in Case 1 on p. 42 the co-ordinates of $\xi_A = \xi$. In other cases $\xi_A = \eta$. We shall now give in each of these cases a formula from which the co-ordinates of η can be calculated. The notation S will be used for the automorphism of $\mathbb{Q}(\zeta_{2f_6})$ induced by $\zeta_{2f_6} \mapsto \zeta_{2f_6}^s$.

In Case 2 on p. 44 Bergström's product formula is of the form

(110) $$6p_0\xi = \tau(\chi_4)(\alpha + i\beta) + \tau(\bar{\chi}_4)(\alpha - i\beta)$$

where $\alpha, \beta \in \mathcal{O}_3$ and the co-ordinates of α and β can be calculated using the equations (106) - (109). In the automorphism S the image of

$$\tau(\chi_4) + \tau(\bar{\chi}_4) = 2\sum_{\chi_4(x)=1} \zeta_{p_0}^x + 2\sum_{\chi_4(x)=-1}(-\zeta_{p_0}^x)$$

is

$$\chi_4(s)^{-1}\left\{2\sum_{\chi_4(x)=\chi_4(s)}\chi_4(x)\zeta_{p_0}^x + 2\sum_{\chi_4(x)=-\chi_4(s)}\chi_4(x)\zeta_{p_0}^x\right\} =$$

$$(-1)^\nu i(\tau(\chi_4) - \tau(\bar{\chi}_4))$$

where $\nu \in \{0,1\}$ is determined by $\chi_4(s) = (-1)^{\nu+1}i$. Further the image of

$$i(\tau(\chi_4) - \tau(\overline{\chi}_4)) = 2 \sum_{\chi_4(x)=i} (-\zeta_{p_0}^x) + 2 \sum_{\chi_4(x)=-i} \zeta_{p_0}^x$$

is

$$2 \sum_{\chi_4(x)=i} (-\zeta_{p_0}^{sx}) + 2 \sum_{\chi_4(x)=-i} \zeta_{p_0}^{sx} = (-1)^{\nu+1}(\tau(\chi_4) + \tau(\overline{\chi}_4)).$$

Hence

$$6p_0\xi' = (-1)^{\nu+1}\{(\tau(\chi_4) + \tau(\overline{\chi}_4))\beta' - i(\tau(\chi_4) - \tau(\overline{\chi}_4))\alpha'\} .$$

Put $\pi(\chi_4,\chi_2) = \sum_{\substack{x,y\in\mathbb{Z} \\ 2\leqslant x\leqslant p_0-1 \\ x+y=1}} \chi_4(x)\chi_2(y) = A + Bi$. Then

$$A^2 + B^2 = p_0,$$

$$\tau(\chi_4)^2 = \chi_2(2)\pi(\chi_4,\chi_2)\tau(\chi_2) = (-1)^{(p_0-1)/4}(A + Bi)\sqrt{p_0},$$

$$\tau(\overline{\chi}_4)^2 = \chi_2(2)\pi(\overline{\chi}_4,\chi_2)\tau(\chi_2) = (-1)^{(p_0-1)/4}(A - Bi)\sqrt{p_0},$$

$$\tau(\chi_4)\tau(\overline{\chi}_4) = \chi_4(-1)p_0 = (-1)^{(p_0-1)/4}p_0$$

[15, p. 466], from which it follows that

(111) $$(\tau(\chi_4) + \tau(\overline{\chi}_4))^2 = (-1)^{(p_0-1)/4}2(A\sqrt{p_0} + p_0),$$

(112) $$(\tau(\chi_4) + \tau(\overline{\chi}_4))(\tau(\chi_4) - \tau(\overline{\chi}_4)) = (-1)^{(p_0-1)/4}2Bi\sqrt{p_0}.$$

Using (111) and (112) we obtain deleting the insignificant sign

$$\eta = \frac{(A\sqrt{p_0} + p_0)\alpha - B\sqrt{p_0}\beta}{(A\sqrt{p_0} + p_0)\beta' + B\sqrt{p_0}\alpha'}$$

whence

(113) $$\eta = \{B(\beta'^2 - \alpha'^2) - 2A\alpha'\beta'\}^{-1}\{B(\alpha\beta' + \alpha'\beta) + A(\beta\beta' - \alpha\alpha')$$
$$- (\alpha\alpha' + \beta\beta')\sqrt{m}\}.$$

In Cases 3 - 6 Bergström's product formula is of the form

(114) $$6f_*\xi = i\sqrt{p_0}\alpha + i\sqrt{p_1}\beta$$

where the co-ordinates of α, $\beta \in \mathcal{O}_3$ can be calculated using the equations (106) - (109), and $p_0 p_1 = f_2$. In the automorphism S $i\sqrt{p_0} \mapsto$ $(-1)^\kappa i\sqrt{p_0}$ and $i\sqrt{p_1} \to (-1)^{\kappa+1} i\sqrt{p_1}$ where $\kappa = 0$ or 1, because $\chi_2(s) = \chi_{2,p_0}(s)\chi_{2,p_1}(s) = -1$. Now

$$\eta = \frac{\sqrt{p_0}\alpha + \sqrt{p_1}\beta}{\sqrt{p_0}\alpha' - \sqrt{p_1}\beta'}$$

from which it follows that

(115) $\eta = (p_0\alpha'^2 - p_1\beta'^2)^{-1}\{p_0\alpha\alpha' + p_1\beta\beta' + (\alpha\beta' + \alpha'\beta)\sqrt{m}\}.$

Correspondingly we obtain η in Cases 9 and 10, in which

(116) $$6f_*\xi = 2\sqrt{2}\alpha + 2\sqrt{p_0}\beta$$

where α, $\beta \in \mathcal{O}_3$. Then

(117) $\eta = (2\alpha'^2 - p_0\beta'^2)^{-1}\{2\alpha\alpha' + p_0\beta\beta' + (\alpha\beta' + \alpha'\beta)\sqrt{m}\}.$

In Cases 7 and 8 Bergström's product formula is of the form

(118) $$6f_*\xi = i2\sqrt{2}\alpha + i2\sqrt{2p_0}\beta$$

where α, $\beta \in \mathcal{O}_3$, $i2\sqrt{2} = \tau(\chi'_{2,8}\chi_{2,4})$ and $i2\sqrt{2p_0} = \tau(\chi'_{2,8}\chi_{2,p_0})$. Since $\chi_2(s) = \chi_{2,4}(s)\chi_{2,p_0}(s) = -1$, we have $\chi'_{2,8}(s)\chi_{2,4}(s) = -\chi'_{2,8}(s)\chi_{2,p_0}(s)$. So in the automorphism S $i2\sqrt{2} \mapsto (-1)^\kappa i2\sqrt{2}$ and $i2\sqrt{2p_0} \mapsto (-1)^{\kappa+1} i2\sqrt{2p_0}$. Hence we have

(119) $\eta = (\alpha'^2 - p_0\beta'^2)^{-1}\{\alpha\alpha' + p_0\beta\beta' + (\alpha\beta' + \alpha'\beta)\sqrt{m}\}.$

On p.10 we have formulas from which we obtain the co-ordinates of the product of two numbers of K_3. The conjugate ω' of an element $\omega = z_0 + z_1\theta + z_2\theta'$ of K_3 is

(120) $$\omega' = \begin{cases} z_0 - z_2 - z_2\theta + (z_1 - z_2)\theta' & \text{if } 3 \nmid f_3 \\ z_0 - z_2\theta + (z_1 - z_2)\theta' & \text{if } 3 \mid f_3. \end{cases}$$

Since $N_{3/1}(\omega) = \omega\omega'\omega''$, we have $\omega^{-1} = N_{3/1}(\omega)^{-1}\omega'\omega''$. Now we have all the needed equations for calculating the co-ordinates of η from (113), (115), (117) or (119). In this connexion we shall also give formulas for calculating the norms $N_{6/2}(\gamma)$ and $N_{6/3}(\gamma)$ of an element $\gamma = \omega_1 + \omega_2\sqrt{m} = x_0 + x_1\theta + x_2\theta' + (y_0 + y_1\theta + y_2\theta')\sqrt{m}$ of K_6. Now

$$N_{6/3}(\gamma) = \gamma\gamma''' = \omega_1^2 - m\omega_2^2,$$

$$N_{6/2}(\gamma) = \gamma\gamma^{iv}\gamma'' = (\omega_1 + \omega_2\sqrt{m})(\omega_1' + \omega_2'\sqrt{m})(\omega_1'' + \omega_2''\sqrt{m}).$$

Using the equations (13), (14) and (120) we obtain the following formulas

Case 1. $3 \nmid f_3$: In this case

$$(121) \quad N_{6/3}(\gamma) = x_0^2 - my_0^2 + \frac{4f_3 - a + 3b - 2}{18}(x_1^2 - my_1^2)$$

$$+ \frac{4f_3 - a - 3b - 2}{18}(x_2^2 - my_2^2)$$

$$+ \frac{2(-f_3 + a - 1)}{9}(x_1x_2 - my_1y_2)$$

$$+ \left\{ \frac{-a + b - 4}{6}(x_1^2 - my_1^2) + 2(x_0x_1 - my_0y_1) \right.$$

$$\left. - \frac{b}{3}(x_2^2 - my_2^2) + \frac{a + b - 2}{3}(x_1x_2 - my_1y_2) \right\}\theta$$

$$+ \left\{ \frac{b}{3}(x_1^2 - my_1^2) - \frac{a + b + 4}{6}(x_2^2 - my_2^2) \right.$$

$$\left. + 2(x_0x_2 - my_0y_2) + \frac{a - b - 2}{3}(x_1x_2 - my_1y_2) \right\}\theta'$$

and

$$(122) \quad N_{6/2}(\gamma) = (x_0 - x_1 - x_2)(x_0^2 + my_0^2) + 2mx_0y_0(y_0 - y_1 - y_2)$$

$$+ \frac{f_3 + 2}{3}\{ x_0(x_1x_2 + my_1y_2) + my_0(x_1y_2 + x_2y_1) \}$$

$$+ \frac{f_3a + 3f_3b - 2}{18}\{ x_2(x_1^2 + my_1^2) + 2mx_1y_1y_2 \}$$

$$+ \frac{1 - f_3}{3}\{ x_0(x_1^2 + x_2^2 + m(y_1^2 + y_2^2)) + 2my_0(x_2y_2 + x_1y_1) \}$$

$$+ \frac{3f_3 - f_3 a - 1}{27} \{ x_1(x_1^2 + 3my_1^2) + x_2(x_2^2 + 3my_2^2) \}$$

$$+ \frac{f_3 a - 3f_3 b - 2}{18} \{ x_1(x_2^2 + my_2^2) + 2mx_2y_1y_2 \}$$

$$+ \Big\{ (y_0 - y_1 - y_2)(x_0^2 + my_0^2) + 2x_0y_0(x_0 - x_1 - x_2)$$

$$+ \frac{f_3 + 2}{3} \{ y_1(x_0x_2 + my_0y_2) + x_1(x_2y_0 + x_0y_2) \}$$

$$+ \frac{f_3 a + 3f_3 b - 2}{18} \{ y_2(x_1^2 + my_1^2) + 2x_1x_2y_1 \}$$

$$+ \frac{1 - f_3}{3} \{ y_0(x_1^2 + x_2^2 + m(y_1^2 + y_2^2)) + 2x_0(x_2y_2 + x_1y_1) \}$$

$$+ \frac{3f_3 - f_3 a - 1}{27} \{ y_1(3x_1^2 + my_1^2) + y_2(3x_2^2 + my_2^2) \}$$

$$+ \frac{f_3 a - 3f_3 b - 2}{18} \{ y_1(x_2^2 + my_2^2) + 2x_1x_2y_2 \} \Big\} \sqrt{m} \ .$$

<u>Case 2. $3|f_3$.</u> In this case

$$(123) \quad N_{6/3}(\gamma) = x_0^2 - my_0^2 + \frac{2f_3}{9} \{ x_1^2 + x_2^2 - x_1x_2 - m(y_1^2 + y_2^2 - y_1y_2) \}$$

$$+ \Big\{ \frac{a + b}{6} (x_1^2 - my_1^2) + 2(x_0x_1 - my_0y_1) - \frac{b}{3}(x_2^2 - my_2^2)$$

$$+ \frac{b - a}{3} (x_1x_2 - my_1y_2) \Big\} \theta$$

$$+ \Big\{ \frac{b}{3} (x_1^2 - my_1^2) + \frac{a - b}{6} (x_2^2 - my_2^2) + 2(x_0x_2 - my_0y_2)$$

$$- \frac{a + b}{3} (x_1x_2 - my_1y_2) \Big\} \theta'$$

and

$$(124) \quad N_{6/2}(\gamma) = x_0(x_0^2 + 3my_0^2) + \frac{f_3}{3} \{ x_0(x_1x_2 - x_1^2 - x_2^2 + m(y_1y_2 - y_1^2$$

$$- y_2^2)) + my_0(x_1y_2 + x_2y_1 - 2x_1y_1 - 2x_2y_2) \}$$

$$- \frac{f_3(a + 3b)}{18} \{ x_1(x_2^2 + my_2^2) + 2mx_2y_1y_2 \}$$

$$- \frac{f_3(a - 3b)}{18} \{ x_2(x_1^2 + my_1^2) + 2mx_1y_1y_2 \}$$

$$+ \frac{f_3 a}{27} \{ x_2(x_2^2 + 3my_2^2) + x_1(x_1^2 + 3my_1^2) \}$$

$$+ \left\{ y_0(3x_0^2 + my_0^2) + \frac{f_3}{3} \{ y_0(x_1x_2 - x_1^2 - x_2^2 + m(y_1y_2 - y_1^2 \right.$$

$$- y_2^2)) + x_0(x_1y_2 + x_2y_1 - 2x_1y_1 - 2x_2y_2) \}$$

$$- \frac{f_3(a + 3b)}{18} \{ y_1(x_2^2 + my_2^2) + 2x_1x_2y_2 \}$$

$$- \frac{f_3(a - 3b)}{18} \{ y_2(x_1^2 + my_1^2) + 2x_1x_2y_1 \}$$

$$\left. + \frac{f_3 a}{27} \{ y_1(3x_1^2 + my_1^2) + y_2(3x_2^2 + my_2^2) \} \right\} \sqrt{m} \ .$$

8. The class number of K_6

Let η_2 and η_3 be the cyclotomic units of K_2 and K_3 respectively as defined in [14, p. 25]. Put $Y_2 = <-1,\eta_2>$, $Y_3 = <-1,\eta_3,\eta_3'>$ and $Y_6 = <-1,\eta_2,\eta_3,\eta_3',\eta,\eta'>$. From Hasse [14, p. 40] we see that the class number h_n of K_n ($n = 2,3,6$) is the index $[U_n : Y_n]$.

<u>Theorem 17.</u> The class number of K_6 is of the form

(125) $$h_6 = h_2 h_3 h_R$$

where h_R is a positive integer which is called the relative class number of K_6.

<u>Proof.</u> From Theorem 9 we obtain $U_6 = <-1,\mu,\tau,\tau',\varepsilon_1,\varepsilon_2>$. We can write

$$\eta_2 = (-1)^{a_{21}} \mu^{a_{22}},$$

$$\eta_3 = (-1)^{a_{31}} \tau^{a_{33}} \tau'^{a_{34}},$$

$$\eta_3' = (-1)^{a_{41}} \tau^{a_{43}} \tau'^{a_{44}},$$

$$\eta = (-1)^{a_{51}} \mu^{a_{52}} \tau^{a_{53}} \tau'^{a_{54}} \varepsilon_1^{a_{55}} \varepsilon_2^{a_{56}},$$

$$\eta' = (-1)^{a_{61}} \mu^{a_{62}} \tau^{a_{63}} \tau'^{a_{64}} \varepsilon_1^{a_{65}} \varepsilon_2^{a_{66}}$$

for some integers a_{ij}. Here $h_2 = \pm a_{22}$ and $h_3 = \pm \begin{vmatrix} a_{33} & a_{34} \\ a_{43} & a_{44} \end{vmatrix}$. Therefore the equation (125) holds with

(126) $$h_R = \pm \begin{vmatrix} a_{55} & a_{56} \\ a_{65} & a_{66} \end{vmatrix}.$$ □

Since $[<-1,\mu,\tau,\tau',\eta,\eta'> : Y_6] = h_2 h_3$, the relative class number

$$h_R = h_6/h_2h_3 = [U_6 : <-1,\mu,\tau,\tau',\eta,\eta'>].$$

If $\xi \in K_6$, then $\eta' = \xi'/\xi'' = N_{6/2}(\xi')N_{6/3}(\xi\xi'')^{-1}\xi.$ So

$$<-1,\mu,\tau,\tau',\eta,\eta'> = <-1,\mu,\tau,\dot{\tau}',\xi_A,\xi_A'>.$$

A generating relative unit ξ_R determines the index $[U_6^* : <-1,\mu,\tau,\tau',$ $\xi_A,\xi_A'>].$ From Theorems 5 - 8 we obtain $[U_6 : U_6^*].$ So we have the value of

(127) $$h_R = [U_6 : U_6^*] \cdot [U_6^* : <-1,\mu,\tau,\tau',\xi_A,\xi_A'>].$$

9. The signature rank of U_6

Let W be a vector space of dimension six over GF(2). The elements of W are denoted by $x = (x_0, \ldots, x_5)$. We define a homomorphism sgn from the multiplicative group of K_6 into the additive group of W as follows: $\text{sgn}(\alpha) = (x_0, \ldots, x_5)$ where $x_i = 0$ if the i'th conjugate $\alpha^{(i)} > 0$ and $x_i = 1$ if $\alpha^{(i)} < 0$. The signature rank Sr of U_6 is then defined as the dimension of the subspace

$$V = \{ \text{sgn}(\omega) \mid \omega \in U_6 \}$$

of W.

Let

$$V_o = \{ \text{sgn}(\omega) \mid \omega \in U_2 U_3 \},$$

i.e. V_o is the subspace of V spanned by the vectors $\text{sgn}(-1)$, $\text{sgn}(\mu)$, $\text{sgn}(\tau)$, $\text{sgn}(\tau')$. In the following theorem a description of V_o is given. We suppress the proof which is simple and straightforward.

<u>Theorem 18.</u> (i) If $N_{2/1}(\mu) = -1$ and τ is not totally positive then $V_o = \{ x \in W \mid x_0 + x_3 = x_1 + x_4 = x_2 + x_5 \}$ and $\dim V_o = 4$.

(ii) If $N_{2/1}(\mu) = 1$ and τ is not totally positive then
$$V_o = \{ x \in W \mid x_0 = x_3, \ x_1 = x_4, \ x_2 = x_5 \} \text{ and } \dim V_o = 3.$$

(iii) If $N_{2/1}(\mu) = -1$ and τ is totally positive then
$$V_o = \{ x \in W \mid x_0 = x_2 = x_4, \ x_1 = x_3 = x_5 \} \text{ and } \dim V_o = 2.$$

(iv) If $N_{2/1}(\mu) = 1$ and τ is totally positive then
$$V_o = \{ x \in W \mid x_0 = x_1 = x_2 = x_3 = x_4 = x_5 \} \text{ and } \dim V_o = 1.$$

Sr depends on $\dim V_o$ as follows:

<u>Theorem 19.</u> We have $Sr = \dim V_o$ or $Sr = \dim V_o + 2$ according as $V = V_o$ or $V \neq V_o$.

<u>Proof.</u> From Theorem 9 it follows that $Sr \leqslant \dim V_o + 2$. Let us suppose that $Sr \neq \dim V_o$, i.e. $V \neq V_o$. Then there exists an element ε in U_6 such that $\mathrm{sgn}(\varepsilon) \notin V_o$. Evidently $\mathrm{sgn}(\varepsilon^{(i)}) \notin V_o$ for every conjugate $\varepsilon^{(i)}$ of ε. Especially $\mathrm{sgn}(\varepsilon'') \notin V_o$. Furthermore $\varepsilon\varepsilon'' = \varepsilon' N_{6/2}(\varepsilon)/N_{6/3}(\varepsilon')$, and so $\mathrm{sgn}(\varepsilon) + \mathrm{sgn}(\varepsilon'') = \mathrm{sgn}(\varepsilon\varepsilon'') \notin V_o$. Thus $Sr = \dim V \geqslant \dim V_o + 2$. $\square$

From Theorems 18 and 19 we see that Sr is odd or even according as $N_{2/1}(\mu) = 1$ or -1.

According to Theorems 5 - 8, $V = V_o$ if the following vectors are in V_o: $\mathrm{sgn}(\xi_A)$, $\mathrm{sgn}(\xi_R)$, $\mathrm{sgn}(\xi_B)$ if ξ_B exists, and $\mathrm{sgn}(\xi_C)$ if ξ_C exists. The following theorem shows that we need not know these vectors to determine Sr if τ is not totally positive.

<u>Theorem 20.</u> Suppose that τ is not totally positive. We have $Sr = \dim V_o$ or $Sr = \dim V_o + 2$ according as $<-1>N_{6/3}(U_6) \neq U_3$ or $<-1>N_{6/3}(U_6) = U_3$.

<u>Remark.</u> It follows readily from results in Section 3 that $<-1>N_{6/3}(U_6) \neq U_3$ if and only if $2|u$, $2|v$ and ξ_C does not exist.

<u>Proof.</u> Put $\overline{U}_3 = <-1>N_{6/3}(U_6)$. Suppose first that $\overline{U}_3 = U_3$. Choose ε so that $N_{6/3}(\varepsilon) = \pm\tau$. Then $\mathrm{sgn}(\varepsilon\varepsilon''') = (x_0, x_1, x_2, x_0, x_1, x_2)$ where x_0, x_1, x_2 are not all equal because τ is not totally positive. From Theorem 18 we see that $\mathrm{sgn}(\varepsilon) \notin V_o$. Suppose next that $\overline{U}_3 \neq U_3$. Then $\overline{U}_3 = <-1, \tau^2, \tau'^2>$. For any $\varepsilon \in U_6$ the relative norms $N_{6/3}(\varepsilon^{(i)})$ ($i = 0, \ldots, 5$) have the same sign so that $\mathrm{sgn}(\varepsilon) = (x_0, \ldots, x_5)$ where $x_0 + x_3 = x_1 + x_4 = x_2 + x_5 = x$, say. Moreover $x = 0$ if $N_{6/1}(\varepsilon) = 1$. Especially is this the case when $N_{2/1}(\mu) = 1$. Thus $\mathrm{sgn}(\varepsilon) \in V_o$. Hence $V = V_o$. Sr is now obtained from Theorem 19. $\square$

10. The computer program

The computer program is constructed for the UNIVAC 1108 system using FORTRAN V programming language. The double precision for real numbers is 18 digits and the greatest allowed absolute value for integers is $2^{35}-1$.

In the beginning of the program we deal with the field K_3. The inputs concerning K_3 are p_0 (or 9 if $3|f_3$), p_1, ... , p_n in (1), the numbers a, b satisfying (2) and (3), $S_{3/1}(\tau)$, $S_{3/1}(\tau^{-1})$, for each p_i a primitive root r_i and the numbers a_i, b_i satisfying $p_i = (a_i^2 + 3b_i^2)/4$, $a_i \equiv 2 \bmod 3$, $b_i \equiv 0 \bmod 3$, $b_i > 0$. The numbers $S_{3/1}(\tau)$, $S_{3/1}(\tau^{-1})$, a, b, a_i, b_i are taken from M.N. Gras's tables [9] bearing in mind the different notation mentioned on p. 6.

If $n = 0$, $\phi = \pi_0$ or 3ρ according as $f_3 = p_0$ or $f_3 = 9$. If $n > 0$, the factors π_i' and the exponent α for $3|f_3$ in the decomposition (83) of ϕ are determined.

Next we compute the values of the character χ_3', defined in (84), for the numbers 1, 2, ... , f_3. The cubic residue symbol $\left(\frac{x}{\pi_i'}\right)_3$ for prime residues x mod p_i can be derived from $\left(\frac{r_i}{\pi_i'}\right)_3$. The value of $\left(\frac{r_i}{\pi_i'}\right)_3$ we calculate as follows. According to (85),

$$r_i^{(p_i-1)/3} \equiv \left(\frac{r_i}{\pi_i'}\right)_3 \bmod \pi_i'.$$

If a rational integer y satisfies the congruence $y \equiv \rho \bmod (a_i \pm b_i\sqrt{-3})/2$, then $p_i|(y-\rho)(a_i \mp b_i\sqrt{-3})/2$ so that $\pm 2yb_i \pm b_i + a_i \equiv 0 \bmod p_i$. Arguing similarly with ρ replaced by ρ^2 we obtain: If $\pi_i' = (a_i \pm b_i\sqrt{-3})/2$ then

$$\left(\frac{r_i}{\pi_i'}\right)_3 = \begin{cases} \rho & \text{if } \pm 2r_i^{(p_i-1)/3} \quad b_i \pm b_i + a_i \equiv 0 \bmod p_i \\ \rho^2 & \text{if } \mp 2r_i^{(p_i-1)/3} \quad b_i \mp b_i + a_i \equiv 0 \bmod p_i. \end{cases}$$

From (86) we have

$$\left(\frac{\rho}{x}\right)_3 = \begin{cases} 1 & \text{if } x \equiv 1,8 \bmod 9 \\ \rho & \text{if } x \equiv 2,7 \bmod 9 \\ \rho^2 & \text{if } x \equiv 4,5 \bmod 9. \end{cases}$$

It is now easy to determine the exponent of ρ in $\chi_3'(x)$ for any prime residue $x \bmod f_3$.

The numerical values of θ, θ', θ'' are calculated in the following way. First we compute θ from (5), which can be written in the form

(128) $$\theta = (-1)^n \sum_{\chi_3'(x)=1} \cos(2\pi x/f_3).$$

At the same time we also compute the number

$$\theta^* = (-1)^n \sum_{\chi_3'(x)=\rho} \cos(2\pi x/f_3).$$

The value of θ obtained from (128) is too inaccurate, especially when f_3 is large. A more precise value we obtain from the minimal polynomial (6) or (9) of θ using Newton's method. As an approximation we use the value obtained from (128). By splitting the numbers into four-digit pieces we calculate θ with an accuracy of 10^{-124}. A precision of a very high order of magnitude is in fact needed later on. The values of θ' and θ'' are calculated from (7) and (8), or (10) and (11) with the same accuracy. If now θ^* is close to θ', then $\chi_3' = \chi_3$; and if θ^* is close to θ'', then $\chi_3' = \overline{\chi}_3$.

Next we determine the values of τ, τ', τ'' and the co-ordinates of τ with respect to the integral basis $\{1,\theta,\theta'\}$. The minimal polynomial of τ is

(129) $$x^3 - S_{3/1}(\tau)x^2 + S_{3/1}(\tau^{-1})x - 1.$$

We first compute a zero of (129) by using Newton's method starting from the first approximation $S_{3/1}(\tau^{-1})^{-1}$. The other zeros are also calculated by Newton's method. The first approximations we get by solving the corresponding quadratic equation. In order to obtain uniquely defined numerical results it is of importance to know the choice of τ among its conjugates. This is done as follows. In the table of M.N. Gras [9] it is always true that

$$|S_{3/1}(\tau^{-1})| > \max \{ |S_{3/1}(\tau)|, |S_{3/1}(\tau) + 2| \} .$$

Using this it is easy to determine the location of the zeros of (129). There is always exactly one zero in the interval $(-1,1)$. The other two zeros are both > 1 if $S_{3/1}(\tau) > 0$, $S_{3/1}(\tau^{-1}) > 0$; they are both < -1 if $S_{3/1}(\tau) < 0$, $S_{3/1}(\tau^{-1}) > 0$; and one of them is > 1 whereas the other is < -1 if $S_{3/1}(\tau^{-1}) < 0$. In the first and third case we take τ to be the greatest zero and in the second case we take it to be the smallest one. Having chosen τ denote the other two zeros by x_2 and x_3. To determine which one is τ' and which one is τ'' consider the system of equations

$$(130) \qquad \begin{cases} \tau = c_0 + c_1\theta + c_2\theta' \\ x_2 = c_0 + c_1\theta' + c_2\theta'' \\ x_3 = c_0 + c_1\theta'' + c_2\theta . \end{cases}$$

If (130) has an integral solution c_0, c_1, c_2, then $x_2 = \tau'$ and $x_3 = \tau''$. If the solution is not integral, then $x_2 = \tau''$ and $x_3 = \tau'$, and we solve the co-ordinates of τ from (130) by interchanging x_2 and x_3.

The inputs concerning the subfield K_2 of K_6 are q_0 (or 4 if $4|f_2$, $8 \nmid f_2$, or 8 if $8|f_2$), $q_1, \ldots, q_k$ in (87), and the co-ordinates z_0, z_1 of the fundamental unit $\mu = z_0 + z_1\sqrt{m}$. The values of the character χ_2, given in (90), will be calculated for each residue class mod f_2. In the sequel we need an accurate value of $\sqrt{m}$ which we compute by Newton's method with the same precision as θ above.

We determine the common prime factors of f_2 and f_3 by comparing the input factors of f_2 and f_3. The product of these factors is f_* and the conductor $f_6 = f_2 f_3 / f_*$, according to (18). From the forms of f_2 and f_3 we then see which one of the cases on pp. 42 - 51 applies.

The set $\alpha = \{a_1, \ldots, a_1\}$ defined on p. 16 will be chosen as follows. First we collect all integers x such that $1 \leqslant x < f_6/2$, $\chi_3(x) = 1$ and $\chi_2(x) = 1$. In this set we only have to replace every even number x by $f_6 - x$ which is odd. From (76) we then compute $B_t (t = 1, 2, \ldots, 2f_6)$. The numbers $B_t^{(i)}$ are split into four pieces, each one containing eight digits. From (77), (78), or (79) we obtain the integers $2A_t$.

Next we determine the co-ordinates of $2f_* \xi_A$ with respect to the field basis $\{1, \theta, \theta', \sqrt{m}, \theta\sqrt{m}, \theta'\sqrt{m}\}$. These co-ordinates are integers, according to Theorem 1. In Case 1 on p. 42 $\xi_A = \xi$ and the computations are carried out by means of the equations (96) and (106) - (109). The numbers are first obtained in eight-digit pieces but for further operations we split the pieces into two parts of four digits. In Cases 2 - 10 $\xi_A = \eta$. We first compute α and β in (110), (114), (116), or (118). The co-ordinates x_i, y_i of $2f_* \eta = x_0 + x_1\theta + x_2\theta' + (y_0 + y_1\theta + y_2\theta')\sqrt{m}$ are obtained from (113), (115), (117), or (119) using the formulas mentioned on pp. 54, 55. It is difficult to make this procedure sufficiently accurate. In some cases we therefore use the following alternative method. We first compute the numerical values of the conjugates of η with a very high precision and then solve x_i, y_i from the system of equations

$$(131) \begin{cases} x_0 + x_1\theta + x_2\theta' = f_*(\eta + \eta'''), & y_0 + y_1\theta + y_2\theta' = f_*(\eta - \eta''')/\sqrt{m}, \\ x_0 + x_1\theta' + x_2\theta'' = f_*(\eta' + \eta^{iv}), & y_0 + y_1\theta' + y_2\theta'' = f_*(\eta' - \eta^{iv})/\sqrt{m}, \\ x_0 + x_1\theta'' + x_2\theta = f_*(\eta'' + \eta^{v}), & y_0 + y_1\theta'' + y_2\theta = f_*(\eta'' - \eta^{v})/\sqrt{m}. \end{cases}$$

The numbers x_i, y_i are again split into four-digit pieces.

The norms $N_{6/2}(\xi_A)$ and $N_{6/3}(\xi_A)$ are computed from (121), (122) or

(123), (124). The numbers u, v and w, defined in (34) and (35), will be determined next. Since $N_{6/3}(\xi_A) = \pm \tau^u \tau'^v$, we have $N_{6/3}(\xi_A') = \pm \tau'^u \tau''^v$. Hence

$$u = \frac{\ln|N_{6/3}(\xi_A)|\ln|\tau''| \ - \ln|N_{6/3}(\xi_A')|\ln|\tau'|}{\ln|\tau|\ln|\tau''| \ - (\ln|\tau'|)^2},$$

$$v = \frac{-\ln|N_{6/3}(\xi_A)|\ln|\tau'| \ + \ln|N_{6/3}(\xi_A')|\ln|\tau|}{\ln|\tau|\ln|\tau''| \ - (\ln|\tau'|)^2}.$$

Using the numerical values of τ, τ', τ'', $N_{6/3}(\xi_A)$ and $N_{6/3}(\xi_A')$ we calculate u and v. Similarly w will be obtained from

$$w = \frac{\ln|N_{6/2}(\xi_A)|}{\ln|\mu|}.$$

In most cases the values obtained in this way are accurate enough.

From (69) we compute the co-ordinates of $2f_* \xi_o$ using the equations (13), (14), (120) and

$$\tau^{-1} = \tau'\tau'', \quad \tau'^{-1} = \tau\tau'', \quad \mu^{-1} = N_{2/1}(\mu)\mu'.$$

Besides the co-ordinates of $2f_* \xi_o$ the numerical values of the conjugates of ξ_o are also computed by using four-digit pieces. As a control we use the relative norms $\xi_o \xi_o'''$, $\xi_o' \xi_o^{iv}$, $\xi_o'' \xi_o^v$ which should be 1, and $\xi_o \xi_o'' \xi_o^{iv}$ which should be ± 1.

If $2|u$, $2|v$ and ξ_o is totally positive or negative, Theorem 3 enables one in many cases to decide that $\pm \xi_o$ is not a square in K_6. If this is not so we calculate

$$\sqrt{|\xi_o|} + \sum_{i=1}^{5} \delta_i \sqrt{|\xi_o^{(i)}|},$$

where $\delta_i \in \{-1,1\}$, with all possible combinations of $\delta_1, \dots, \delta_5$. The square roots are computed by Newton's method using four-digit pieces. If none of the results is an integer, then $\pm \xi_o$ is not a square in K_6. Otherwise we solve the system of equations obtained from (131) by substituting $\sqrt{|\xi_o|}$ for η and $\delta_i \sqrt{|\xi_o^{(i)}|}$ for $\eta^{(i)}$ (i = 1, ... ,5). If the system has a solution in integers, then $\pm\xi_0$ is a square in K_6. In such a case we proceed in the manner explained on p. 29.

Next we calculate $\mathcal{M}(\xi_1) = \sum_{i=0}^{5} \xi_1^{(i)^2}/6$, M_{min} from (70), and K_{max} from (73). Then we calculate the right side of (75) for all integers K, L such that $0 \leqslant K, L \leqslant K_{max}$, $K > 0$ and $K^2 + KL + L^2$ odd. For K = L = 1 the integral part of the resulting number may become too large. In this case therefore we calculate the cube roots by Newton's method using four-digit pieces. For the other values of K and L we have not done so. If in each case the result is not an integer, then ξ_1 is a generating relative unit. If for some K, L the result is an integer, then we find out whether $\xi_1^{L/(K^2+KL+L^2)}\xi_1^{K/(K^2+KL+L^2)}$ belongs to K_6 by solving the system of equations obtained from (131) by an obvious modification. In this way we determine ξ_R.

The structure of U_6 is obtained from Theorems 4 - 8. If $3|w$ we have to find out whether one of the equations (37) has a solution ξ_B in K_6. Consider e.g. the first equation. We compute $\sum_{i=0}^{5} (\mu^{(i)}\xi_R^{(i)}\xi_R^{(i+1)})^{1/3}$ and proceed as before. Here again we get the cube roots by Newton's method using four-digit pieces. If $2|u$ and $2|v$, then we find out whether one of the equations (42) has a solution ξ_C in K_6 in the same way as we solved the existence or nonexistence of $\sqrt{|\xi_0|}$ in K_6.

The relative class number h_R is determined from (127). The signature rank Sr of U_6 is obtained from Theorems 18 and 20 if τ is not totally positive. If τ is totally positive, then we calculate the signatures of the generators of U_6, and determine Sr by Theorems 18 and 19.

11. Numerical results

All the real cyclic sextic fields with conductor $f_6 \leqslant 2021$ have been under consideration with 12 exceptions. In six cases the required information in the table of M.N. Gras [9] is missing, and in the other six cases the numbers appearing are too large to be handled by the program. The smallest value of f_6 for which such an exceptional case appears is 997. In the following table the fields are arranged lexicographically according to increasing values of f_6, f_2, f_3, b. Also in the 12 cases mentioned above the known quantities have been written down in their right positions. So we have a complete list of real cyclic sextic fields with conductor $\leqslant 2021$.

The meaning of the numbers and letters in the table is as follows:

$$
\begin{array}{cccc}
f_6 & k & Sr & h_R \\
f_2 & code1 & code2 & h_2 \\
f_3 & a & b & h_3 \\
 & u & v & w
\end{array}
\left\{
\begin{array}{l}
\text{the co-ordinates of } \xi_A \\
\text{multiplied by } k \\
\text{the co-ordinates of } \xi_R \\
\text{multiplied by } k.
\end{array}
\right.
$$

The co-ordinates of ξ_A and ξ_R are taken with respect to the field basis $\{1,\theta,\theta',\sqrt{m},\theta\sqrt{m},\theta'\sqrt{m}\}$ and k is a suitable multiplier which makes these co-ordinates integral. In fact we have chosen $k = 2f_*$ if f_2 is odd and $k = f_*$ if f_2 is even, an allowable choice by Theorem 1. In some cases this multiplier is too large so that the co-ordinates may have a non-trivial factor in common with k. Further

$$
code1 = \begin{cases} d \text{ if } \xi_A = \xi \\ n \text{ if } \xi_A = \eta, \end{cases}
\qquad
code2 = \begin{cases} o \text{ if } \xi_R = \xi_o \\ r \text{ if } \xi_R \neq \xi_o. \end{cases}
$$

In the case $h_R = 7$ the letter r is followed by 1 or 2 depending on whether $\xi_R^7 = \xi_0^2 \xi_0'$ or $\xi_R^7 = \xi_0 \xi_0'^2$.

In a few cases the co-ordinates of ξ_R are not included in the table because of insufficient typing space. All the calculations, however, have been completed in these cases, and the co-ordinates of ξ_R are the only quantities missing from the table.

If ξ_B or ξ_C exists, then its co-ordinates multiplied by k are written under the co-ordinates of $k\xi_R$. The code beside these co-ordinates shows what unit is in question. For each code the corresponding equation having the unit as a root is the following one:

$$\begin{array}{llll}
b1 & x^3 = \mu\xi_R\xi_R', & b2 & x^3 = \mu^{-1}\xi_R\xi_R', \\
c2 & x^2 = \tau\xi_R, & c\text{-}2 & x^2 = -\tau\xi_R, \\
c3 & x^2 = \tau'\xi_R, & c\text{-}3 & x^2 = -\tau'\xi_R, \\
c4 & x^2 = \tau\tau'\xi_R, & c\text{-}4 & x^2 = -\tau\tau'\xi_R.
\end{array}$$

If $h_R = 1$, then it follows e.g. from (127) that $\{\mu,\tau,\tau',\xi_A,\xi_A'\}$ forms a system of fundamental units of K_6. We have discovered altogether 130 fields with $h_R > 1$. In what follows we shall divide these fields into eight different types. For each type we give the number of cases (= no.) and a suitable pair of units ε_1, ε_2 such that $\{\mu,\tau,\tau',\varepsilon_1,\varepsilon_2\}$ is a system of fundamental units (cf. Theorem 9). The choice depends on trivial but somewhat lengthy computations based on the observations made after the equation (25).

1) $\underline{h_R = 4, \xi_R^2 = \pm\xi_0}$, no. = 11; $\varepsilon_1 = \xi_R/\xi_A$, $\varepsilon_2 = \xi_R'/\xi_A'$.

2) $\underline{h_R = 9, \xi_R^3 = \xi_0}$, no. = 1, viz. $(f_6,f_2,f_3,a,b) = (1548,172,387, -39,3)$; $\varepsilon_1 = \xi_R/\xi_A$, $\varepsilon_2 = \xi_R'/\xi_A'$. We note that this choice does not work for $2|u$, $2|v$. We further note that for a field of this type always $3|w$. This can be seen easily in the following way. Suppose that $3 \nmid w$. Then

$$(\xi_R \xi_R')^{3e} = \mu^{2w} \tau^{3v} \tau'^{-3u+3v} \xi_A'^{6}$$

where e = 1 or 2, a contradiction by Theorem 2.

3) $h_R = 3$, $\xi_R^3 = \xi_0 \xi_0'^2$ no. = 12; $\varepsilon_1 = \xi_R/\xi_A$, $\varepsilon_2 = \xi_R'/\xi_A'$. Again we have $3|w$. Suppose namely that $3\nmid w$. Then we get a similar contradiction as in the preceding case, viz.

$$\xi_R^{3e} = \mu^{2w} \tau^{3v} \tau'^{-3u+3v} \xi_A'^{6}$$

where e = 1 or 2.

4) $h_R = 7$, $\xi_R^7 = \xi_0^2 \xi_0'$ or $\xi_0 \xi_0'^2$, no. = 23;

$$\varepsilon_1, \varepsilon_2 = \begin{cases} \xi_R, \xi_R' & \text{if } 3|w, \ 2|u, \ 2|v \\ \xi_R/\xi_A, \ \xi_R/\xi_A' & \text{if } 3|w, \text{ and } 2\nmid u \text{ or } 2\nmid v \\ \xi_R/\xi_A', \ \xi_R'/\xi_A'' & \text{if } 3\nmid w. \end{cases}$$

5) $h_R = 3$, $\xi_R = \xi_0$, ξ_B exists, no. = 28; $\varepsilon_1 = \xi_B/\xi_A$, $\varepsilon_2 = \xi_B'/\xi_A'$.

6) $h_R = 4$, $\xi_R = \xi_0$, ξ_C exists, no. = 53; $\varepsilon_1 = \xi_C/\xi_A$, $\varepsilon_2 = \xi_C'/\xi_A'$.

7) $h_R = 12$, $\xi_R = \xi_0$, ξ_B and ξ_C both exist, no. = 1, viz. $(f_6, f_2, f_3, a, b) = (995, 5, 199, 11, 15)$; $\varepsilon_1 = \xi_B/\xi_C$, $\varepsilon_2 = \xi_C'$.

8) $h_R = 16$, $\xi_R^2 = \pm\xi_0$, ξ_C exists, no. = 1, viz. $(f_6, f_2, f_3, a, b) = (1143, 381, 1143, -3, 39)$; $\varepsilon_1 = \xi_C$, $\varepsilon_2 = \xi_C'/\xi_A'$. The choice does not work for $3|w$.

In the third and fourth case above the ξ_R in the table is the same as the ξ_R' in (74).

In Section 9 we observed that it is more difficult to determine the signature rank of U_6 when τ is totally positive. It is therefore of importance to be able to recognize these cases. There are exactly three of them, viz. $(f_6, f_2, f_3, a, b) = (703, 37, 703, -25, 27)$, $(711, 237, 711, -12, 30)$, $(1009, 1009, 1009, -43, 27)$.

We give three examples of how to use the table.

Example 1. $(f_6, f_2, f_3, a, b) = (728, 104, 91, 11, 9)$, page 105.

$Sr = 6$, $h_R = 1$, $h_2 = 2$, $h_3 = 3$,

$\xi_A = \xi = \frac{1}{13}(-143 + 39\,\theta' + (-16 + 6\,\theta + 11\,\theta')\sqrt{26})$,

$\xi_R = \xi_o = \frac{1}{13}(-285831 + 44720\,\theta + 129324\,\theta' +$

$$(-61624 + 7106\,\theta + 24238\,\theta')\sqrt{26}),$$

$N_{6/3}(\xi_A) = \pm\tau\tau'^2$, $\quad N_{6/2}(\xi_A) = \pm\mu^{-2}$,

$h_6 = 6$, $\quad U_6 = \langle -1,\mu,\tau,\tau',\xi_A,\xi_A' \rangle$.

Example 2. $(f_6,f_2,f_3,a,b) = (995,5,199,11,15)$, page 123.

$Sr = 6$, $h_R = 12$, $h_2 = h_3 = 1$,

$\xi = \xi_A = \xi_o = \xi_R = \frac{1}{2}(277 - 20\,\theta - 150\,\theta' + (-303 + 36\,\theta + 50\,\theta')\sqrt{5})$,

$N_{6/3}(\xi) = |N_{6/2}(\xi)| = 1$,

$\xi_B = \frac{1}{2}(31 - 5\,\theta - 10\,\theta' + (-17 + \theta + 4\,\theta')\sqrt{5})$, $\quad \xi_B^{\,3} = \mu^{-1}\xi\xi'$,

$\xi_C = \frac{1}{2}(131 - 14\,\theta - 36\,\theta' + (-59 + 6\,\theta + 16\,\theta')\sqrt{5})$, $\quad \xi_C^{\,2} = \tau'\xi$,

$h_6 = 12$, $\quad U_6 = \langle -1,\mu,\tau,\tau',\xi_B/\xi_C,\xi_C' \rangle$.

Example 3. $(f_6,f_2,f_3,a,b) = (556,556,139,23,3)$, page 96.

$Sr = 5$, $h_R = 7$, $h_2 = h_3 = 1$,

$\xi_A = \eta = \frac{1}{139}(1610454 - 420614\,\theta + 35723\,\theta' +$

$$(-136597 + 35676\,\theta - 3030\,\theta')\sqrt{139}),$$

$\xi_R = \frac{1}{139}(139 - 278\,\theta + 278\,\theta' + (-170 + 26\,\theta + 20\,\theta')\sqrt{139})$,

$N_{6/3}(\xi_A) = \pm\tau^2\tau'$, $\quad N_{6/2}(\xi_A) = \pm\mu^{-1}$,

$h_6 = 7$, $\xi_R^{\,7} = \xi_o\xi_o'^2$, $\quad U_6 = \langle -1,\mu,\tau,\tau', \xi_R/\xi_A', \xi_R/\xi_A'' \rangle$.

```
13   26  6  1        -13        -13        -13
13    n  o  1          5         -1          3
13    5  3  1        -13         26         26
     -1 -1 -1          3          2         -6

21   14  5  1         21         -7         14
21    n  o  1         -5          1         -2
 7   -1  3  1         -7         42         42
      2  1 -1         -3          2         10

28    7  5  1         -7          7         -7
28    n  o  1          3         -2          4
 7   -1  3  1         -7        -14        -28
      1  2 -1          6         -4         -6

35    2  6  1         -1         -2         -2
 5    d  o  1          1          0          0
 7   -1  3  1         27         30         10
      0  1 -1        -17        -10         -6

36    3  5  1          3         -3          0
12    n  o  1          0          1         -1
, 9  -3  3  1          3         18         18
     -2 -1 -1        -12         -6          0

37   74  6  1       -148        -37        -37
37    n  o  1         28          3          7
37   11  3  1        222        296        148
     -1 -1 -1       -112        -12        -28

45    2  6  1          1          0          2
 5    d  o  1         -1          0          0
 9   -3  3  1        -13        -30        -30
     -1  0 -1         15         10          2

56    1  6  1          0          1          0
 8    d  o  1         -1          0          0
 7   -1  3  1         -3         -4          0
      0  1  0          4          2          2

56    7  5  1         14          7          7
56    n  o  1         -4         -2         -3
 7   -1  3  1       -189       -112        -84
      1  2 -1         34         38          8

57   38  5  1        247         95         38
57    n  o  1        -31        -13         -4
19   -7  3  1       -304        114        -57
     -1  1 -1        -30         -4        -29

61  122  6  1        183          0          0
61    n  o  1         21         -8         10
61   -1  9  1      -8113       3294       5490
     -1 -1 -1       1335       -282       -654
```

63	6	5	1	15	−12	6
21	n	o	1	−3	2	−2
9	−3	3	1	−183	126	0
	−2	−1	−1	21	−18	24
63	42	5	1	−42	21	21
21	n	o	1	0	1	5
63	15	3	3	−1281	−756	126
	2	1	−1	525	156	66
63	42	5	1	399	−42	−168
21	n	o	1	−105	8	40
63	−12	6	3	42	63	63
	1	2	−1	42	9	3
65	2	4	1	3	1	2
5	d	o	1	1	1	0
13	5	3	1	−1	−8	−2
	0	0	−1	−5	0	−2
65	26	4	1	−26	13	26
65	d	o	2	2	−3	−4
13	5	3	1	52	26	39
	0	0	−1	6	4	1
72	1	6	1	0	−1	0
8	d	o	1	1	0	0
9	−3	3	1	25	−24	12
	−1	−1	−1	24	−14	8
72	3	5	1	−9	6	−3
24	n	o	1	3	−2	2
9	−3	3	1	−69	36	72
	−2	−1	−1	−24	6	30
73	146	6	1	2336	−219	584
73	n	o	1	−282	27	−70
73	−7	9	1	2482	1460	219
	−1	0	−1	−582	−126	−87
76	19	5	1	−171	−38	−57
76	n	o	1	39	9	13
19	−7	3	1	57	76	38
	−1	1	−1	−26	−6	4
77	14	5	1	70	−35	42
77	n	o	1	−8	3	−6
7	−1	3	1	−987	2310	3850
	2	1	−1	−107	230	450
84	1	5	1	−1	−1	−1
12	d	o	1	1	0	0
7	−1	3	1	−89	−66	−36
	1	2	−1	46	40	14

91	2	4	1	3	-2	1
13	d	o	1	-1	0	-1
7	-1	3	1	1	-2	-10
	0	0	-1	1	-2	-2

91	26	6	1	-13	0	0
13	d	o	1	-25	-8	-2
91	-16	6	3	-13	-65	-52
	-1	0	-1	-57	1	10

91	26	6	1	0	-13	-26
13	d	o	1	20	-1	-4
91	11	9	3	-23777	2886	9698
	0	1	-1	-6321	942	2754

93	62	5	1	279	-62	124
93	n	o	1	-13	2	-10
31	-4	6	1	155	279	0
	-2	-1	-1	-73	-15	-18

95	2	6	1	7	1	3
5	d	o	1	-1	1	-1
19	-7	3	1	-33	-10	-10
	-1	0	0	-3	-6	2

97	194	6	1	-5432	194	-1164
97	n	o	1	550	-20	118
97	-19	3	1	194	-388	388
	0	-1	-1	-284	-32	-44

99	6	5	1	0	12	3
33	n	o	1	0	-2	-1
9	-3	3	1	105	198	99
	-2	-1	-1	-33	-24	-21

104	1	6	1	-2	0	-1
8	d	o	1	0	-1	0
13	5	3	1	-23	36	-4
	-1	-1	-1	34	-14	12

104	13	6	1	26	0	-13
104	d	o	2	-2	3	4
13	5	3	1	6565	4212	3276
	-1	-1	-1	1374	786	528

105	14	5	1	28	7	-14
105	d	o	2	-2	-1	2
7	-1	3	1	-301	105	-315
	1	2	-1	-31	23	-11

109	218	6	1	17985	-3052	2834
109	n	o	1	-1701	288	-268
109	2	12	1	109	-109	-218
	0	-1	-1	67	-5	-12

117	2	6	1	-3	-1	3
13	d	o	1	-1	1	1
9	-3	3	1	-37	26	52
	-1	0	0	-13	2	12
117	26	6	1	-78	0	26
13	d	o	1	26	-2	-8
117	-21	3	3	26	-52	52
	0	-1	0	104	-16	-12
117	26	6	1	78	13	0
13	d	o	1	-26	-11	-6
117	6	12	3	26	-13	-26
	0	1	0	26	-3	-4
119	2	6	1	1	2	-2
17	d	o	1	-1	0	0
7	-1	3	1	87	-17	153
	1	1	-1	31	-25	-1
124	31	3	1	279	-124	93
124	n	o	1	-50	22	-17
31	-4	6	1	0	-31	-62
	0	0	-1	17	-5	-6
129	86	5	1	-23048	8944	-3870
129	n	o	1	2030	-788	342
43	8	6	1	-1236766	-490716	-335400
	-2	-1	-1	107848	43376	30424
133	14	5	1	-63	7	-105
133	n	o	1	5	-1	9
7	-1	3	1	-3577	-532	-1862
	1	2	-1	-11	-212	46
133	38	3	1	-133	190	247
133	n	o	1	13	-16	-21
19	-7	3	1	-323	-76	76
	0	0	-1	-17	-20	-12
133	266	5	1	-81928	-20216	-15029
133	n	o	1	7094	1754	1307
133	17	9	3	52535	-324786	-66766
	2	1	-1	119157	186	15342
133	266	5	1	-266	-1330	-1197
133	n	o	1	-52	122	121
133	-10	12	3	133	-133	-266
	-2	-1	-1	81	-11	-12
140	7	5	1	-21	0	7
140	d	o	2	2	1	-2
7	-1	3	1	3087	-1470	2870
	-1	1	-1	578	-362	276

152	1	6	1	1	1	-1
8	d	o	1	2	1	1
19	-7	3	1	-1291	-912	-76
	-1	0	-1	-1236	-598	-222
152	19	5	1	0	19	38
152	n	o	1	4	-2	-5
19	-7	3	1	8151	6916	3040
	-1	1	-1	1758	660	-98
153	2	4	1	-3	-5	-3
17	d	o	1	1	1	1
9	-3	3	1	-5	9	-3
	0	0	-1	-3	1	-1
155	2	6	1	-6	1	-3
5	d	o	1	4	-1	1
31	-4	6	1	-8	5	5
	-1	0	0	-2	1	3
156	1	5	1	-3	0	-1
12	d	o	1	0	1	0
13	5	3	1	-5	6	0
	-2	-1	0	-4	2	-2
156	13	5	1	-39	0	13
156	d	o	2	2	-3	-4
13	5	3	1	-14807	-9126	-6942
	-2	-1	-1	-2426	-1444	-998
157	314	6	1	-64998	12089	-5966
157	n	o	1	5124	-955	470
157	14	12	1	314	-157	157
	-1	-1	-1	-100	13	1
161	14	5	1	-189	126	280
161	n	o	1	15	-10	-22
7	-1	3	1	14	-322	-805
	1	2	-1	-26	36	47
168	1	5	1	2	-1	-1
24	d	o	1	0	0	1
7	-1	3	1	-11	-48	-156
	1	2	-1	-30	38	40
168	7	5	1	28	7	7
168	d	o	2	-4	-2	-3
7	-1	3	1	-7301	-6048	-2940
	1	2	-1	1142	886	356
171	6	3	1	-21	-39	-30
57	n	o	1	3	5	4
9	-3	3	1	-21	9	-18
	0	0	-1	-3	3	0

171	114	5	1	12882	-1710	-570
57	n	o	1	228	-178	-398
171	24	6	3	9903066	2860488	623124
	2	1	-1	-1563168	-326256	-20880
171	114	5	1	-114	0	114
57	n	o	1	0	2	-14
171	-3	15	3	114	-684	0
	2	1	-1	456	24	60
172	43	3	1	-645	215	-172
172	n	o	1	98	-33	26
43	8	6	1	43	-43	43
	0	0	-1	-26	7	1
180	3	5	1	-9	3	0
60	d	o	2	0	1	-1
9	-3	3	1	-717	450	90
	-2	-1	-1	-60	66	-120
181	362	6	1	2715	-181	724
181	n	o	1	-11	-7	-26
181	-7	15	1	-82717	14842	30770
	1	1	-1	8241	-1354	-1978
185	2	6	1	5	-3	-3
5	d	o	1	-1	-1	1
37	11	3	1	1977	5570	10160
	-1	-1	-1	9357	1602	-1096
185	74	6	1	-925	444	-111
185	d	o	2	65	-34	7
37	11	3	1	-21756	11285	-1480
	-1	-1	-1	1590	-789	194
193	386	6	1	-59637	-4246	-7913
193	n	o	1	4281	308	569
193	23	9	1	280236	-14475	-77200
	0	1	-1	-21900	1245	5430
201	134	3	1	22914	6365	4422
201	n	o	1	-1616	-449	-312
67	5	9	1	1876	469	268
	0	0	-1	-90	-43	-26
203	2	4	1	-15	-6	-8
29	d	o	1	1	2	0
7	-1	3	1	-15	-6	-8
	0	0	0	1	2	0
207	6	5	1	-33	15	-6
69	n	o	1	3	-3	0
9	-3	3	1	-615	-4968	-1242
	-1	1	-1	-621	-192	-366

209	38	3	1	−114	608	969
209	n	o	1	8	−42	−67
19	−7	3	1	−95	−57	19
	0	0	−1	−9	−5	−3
215	2	6	1	−1	0	3
5	d	o	1	−1	0	1
43	8	6	1	−693	−185	−180
	−1	−1	−1	205	123	62
217	14	5	1	−6349	2961	−4375
217	n	o	1	431	−201	297
7	−1	3	1	−637	1085	1519
	2	1	−1	−13	53	125
217	62	5	1	62	−4154	682
217	n	o	1	−4	282	−46
31	−4	6	1	10245066	5539576	1785476
	−2	−1	−1	−814216	−348416	−163800
217	434	5	1	−746914	11284	170996
217	n	o	1	50704	−766	−11608
217	29	3	3	−434	−868	−1736
	−1	−2	−1	−592	−52	12
217	434	5	1	−999502	−208320	−37541
217	n	o	1	67842	14142	2547
217	−25	9	3	−157759	−1953	−48825
	1	−1	−1	−17649	1311	−1527
221	2	6	1	1	2	2
17	d	o	1	1	0	0
13	5	3	1	19	17	17
	0	−1	0	7	3	1
221	26	5	1	−650	468	−195
221	d	o	2	40	−34	11
13	5	3	1	−1462331	1151852	−456586
	0	−1	−1	100021	−78252	28030
228	1	5	1	1	2	−1
12	d	o	1	3	1	1
19	−7	3	1	451	516	−78
	−1	1	−1	666	238	164
229	458	6	1	2564113	−141522	404643
229	n	o	3	−169297	9344	−26717
229	−22	12	1	229	−229	−458
	1	1	−3	143	−17	−12
231	2	5	1	12	5	5
33	d	o	1	−2	−1	−1
7	−1	3	1	662	561	264
	2	1	−1	−120	−91	−38

237	158	5	1	4345	237	790
237	n	o	1	-285	-15	-50
79	17	3	1	-621967	-533250	-865050
	1	2	-1	177455	28570	170
241	482	6	1	50610	-23136	6507
241	n	o	1	-3258	1490	-419
241	17	15	1	-157855	10845	36873
	-1	0	-1	9729	-897	-2451
247	2	6	1	-5	2	-1
13	d	o	1	1	0	1
19	-7	3	1	-115	182	208
	-1	-1	-1	-51	38	56
247	26	6	1	78	13	13
13	d	o	1	-8	3	-1
247	-31	3	3	-188799	32318	27352
	-1	0	-1	-31295	8718	11264
247	26	6	1	1092	195	78
13	d	o	1	-86	-23	-14
247	-4	18	3	-85384	11583	21177
	-1	-1	-1	-23682	3099	5883
248	1	4	1	-7	-2	-2
8	d	o	1	2	2	0
31	-4	6	1	-7	-2	-2
	0	0	0	2	2	0
248	31	5	4	651	-124	248
248	n	o	1	-86	18	-28
31	-4	6	1	-1705	868	1612
	0	0	-1	-220	134	198
			c3	-155	0	-62
				22	-1	5
252	3	5	1	-6	-3	0
12	d	o	1	-3	-2	-1
63	15	3	3	-3021	90	-1188
	-1	1	-1	1932	-126	606
252	3	3	1	-6	3	3
12	d	o	1	-3	-1	1
63	-12	6	3	3	-18	-27
	0	0	-1	42	6	-3
252	1	5	1	1	-3	0
28	d	o	1	1	0	1
9	-3	3	1	1	70	14
	-1	-2	0	-28	-6	-16
252	7	5	1	7	14	7
28	d	o	1	-14	-1	2
63	15	3	3	7	28	14
	-1	1	0	-28	-2	4

```
252    7   3   1          -7              7              -7
 28    d   o   1          14             -3              -1
 63  -12   6   3          -7              7              -7
       0   0   0          14             -3              -1

259    2   6   1          -8             -1              -4
 37    d   o   1           0             -1               0
  7   -1   3   1         -35            -74               0
       0   1   0         -17             -6              -8

259   74   6   1         296             37               0
 37    d   o   1          16              7               4
259  -19  15   3        -333             74              74
      -1  -1   0          55             -6             -14

259   74   6   1       -3293           -703            -444
 37    d   o   1        -667           -139             -86
259    8  18   3       -3589            407            -444
       0   1   0         651            -87              42

261    2   6   1          -7              3              -2
 29    d   o   1           1             -1               0
  9   -3   3   1        1307          -1044             522
      -1  -1  -1        -261            180            -102

268   67   5   1      364547         -92862           64454
268    n   o   1      -44537          11345           -7874
 67    5   9   1     -267933          61774           67938
      -2  -1  -1       -7434           1098           12780

273    2   5   1          32             -9              10
 21    d   o   1          -2              5               0
 13    5   3   1      117161        -122094           31584
      -1  -2  -1      -35021          20918          -11048

273   14   5   1        -182            -49             -14
 21    d   o   1         -40            -13              -2
 91  -16   6   3        1484            525              63
      -2  -1  -1         410            107              31

273   14   3   1          56              7               7
 21    d   o   1          -6             -3              -1
 91   11   9   3         217            -42              14
       0   0  -1          29            -10               6

273   14   3   1          98            168              21
273    d   o   2          -6            -10              -1
  7   -1   3   1        -112           -126            -105
       0   0  -1           8              4              -1

273   26   5   1        -143            117            -130
273    d   o   2           9             -7               8
 13    5   3   1       -1885          -1365            -819
      -1  -2  -1        -133            -67             -59
```

273	182	5	1	728	910	728
273	d	o	2	-34	-66	-36
91	-16	6	3	-2232150466	783777540	975333996
	-2	-1	-1	137625416	-46672648	-58875920
273	182	3	1	2366	-455	-182
273	d	o	2	-138	29	12
91	11	9	3	3094	0	455
	0	0	-1	-12	50	5
277	554	4	1	-6934141	1096366	-333508
277	n	o	1	416745	-65892	20044
277	26	12	4	-554	277	-277
	2	2	-1	176	-19	-7
279	6	5	1	48	33	39
93	n	o	1	-6	-3	-3
9	-3	3	1	-161535	-177444	-112158
	-1	1	-1	15903	18732	12606
279	186	5	1	-5952	-1023	-93
93	n	o	1	-558	-111	-3
279	33	3	3	3710607	-108810	703080
	1	-1	-1	-457839	-2190	-73680
279	186	5	1	-5952	930	-651
93	n	o	1	-744	96	-15
279	-21	15	3	-233337	15066	45198
	1	2	-1	22599	-1278	-4770
280	1	6	1	-7	-5	-1
40	d	o	2	2	2	1
7	-1	3	1	-3219	-2660	-1120
	-1	1	-2	1056	822	386
280	7	5	1	-49	35	-7
280	d	o	2	6	-4	1
7	-1	3	1	707	-420	560
	-1	1	-1	-96	50	-58
285	38	5	1	-247	-38	-95
285	d	o	2	15	2	5
19	-7	3	1	2603	7410	5700
	-1	1	-1	-579	14	244
287	2	4	1	2	-5	-12
41	d	o	1	0	1	2
7	-1	3	1	22	19	4
	0	0	-1	4	3	2
296	1	6	1	9	-3	1
8	d	o	1	-4	3	0
37	11	3	1	-15271	7508	-1524
	0	-1	-1	10430	-5464	938

```
296   37   6   1          185          37          37
296    d   o   2          -16          -7          -4
 37   11   3   1       -22755       10804       -2516
       0  -1  -1        -2478        1344        -194

301   14   3   1         -539        -329         -70
301    n   o   1           31          19           4
  7   -1   3   1           91         -42          28
       0   0  -1            3          -2           4

301   86   5   1         2537       -1720         344
301    n   o   1         -159         104         -22
 43    8   6   1         -215         301       -1204
      -1   1  -1         -171          51          38

301  602   5   1       124012       51471      -14147
301    n   o   1        -7138       -2967         817
301  -31   9   3   -549960411   -95338138   -15686916
       2   1  -1     31639443     5507010      920028

301  602   5   1      2281580     -324779      130333
301    n   o   1      -131514       18719       -7513
301   23  15   3     -2302951      542402     -839188
      -1   1  -1       436383      -49674      -15844

305    2   6   1          -34           9          -8
  5    d   o   1           18          -3           4
 61   -1   9   1           77         -30          20
      -1  -1   0          -57           6         -12

305  122   5   1          549          61         183
305    d   o   2          -25          -5          -9
 61   -1   9   1        -4758        1525        2440
      -1  -1  -1          234         -63        -150

308    1   5   1           -1          -2          -6
 44    d   o   1           -1           1           1
  7   -1   3   1          -21          44          66
       1   2   0            4         -10         -22

309  206   3   1        -5768        1957        2575
309    n   o   1          326        -111        -147
103  -13   9   1         2781       -1236       -1648
       0   0  -1         -265          52          80

312    1   5   1           -3           3           5
 24    d   o   1            1           0          -2
 13    5   3   1        27709      -17928      -54636
      -2  -1  -1       -16304        4298       20110

312   13   5   1          195         117          91
312    d   o   2          -21         -14         -10
 13    5   3   1        -3575        2808       -1404
      -2  -1  -1         -436         342         -90
```

313	626	6	1	9503306	-1661404	91396
313	n	o	1	-537158	93908	-5166
313	35	3	7	-626	-1252	-2504
	-1	2	-1	852	64	-12
315	2	4	1	-4	-7	-3
5	d	o	1	8	1	-1
63	15	3	3	-229	-96	-18
	0	0	-2	105	40	2
315	2	6	1	25	-3	-11
5	d	o	1	13	-1	-5
63	-12	6	3	-418	60	195
	1	2	-2	-210	14	79
315	6	5	1	-27	9	-15
105	d	o	2	3	-1	1
9	-3	3	1	321	945	945
	-2	-1	-1	-105	-63	-9
315	42	3	1	189	-63	-42
105	d	o	2	-21	5	4
63	15	3	3	-399	-126	-63
	0	0	-1	-21	-12	3
315	42	5	1	630	-168	-126
105	d	o	2	0	4	-22
63	-12	6	3	-452906958	111100500	31689000
	1	2	-1	14910000	-7829760	9172200
316	79	5	1	27887	-10507	790
316	n	o	3	-3138	1182	-89
79	17	3	1	237	316	158
	1	2	-3	114	6	20
323	2	6	1	6	-2	3
17	d	o	1	-2	0	-1
19	-7	3	1	-32	0	-17
	-1	0	0	6	-2	3
329	14	5	1	182	-2576	-3990
329	n	o	1	-10	142	220
7	-1	3	1	2646	-9212	-19740
	1	2	-1	-356	620	944
333	2	4	1	-13	2	-8
37	d	o	1	-1	2	0
9	-3	3	1	-13	2	-8
	0	0	0	-1	2	0
333	74	6	1	-740	-37	37
37	d	o	1	-296	-35	-15
333	-30	12	3	74	-37	37
	-1	-1	0	-74	7	3

333	74	6	1	4847	777	333
37	d	o	1	629	143	87
333	-3	21	3	-69745	-7030	-518
	1	1	0	-4181	-1814	-1470
335	2	4	1	-5	1	-2
5	d	o	1	3	-1	0
67	5	9	1	-17	4	18
	0	0	-1	-19	4	6
337	674	6	1	-12165700	1254651	-1059528
337	n	o	1	662706	-68345	57716
337	5	21	1	-41451	-9099	-3707
	1	1	-1	3043	415	289
341	62	5	1	4960	-1147	1798
341	n	o	1	-270	63	-98
31	-4	6	1	-10168	-8525	-1023
	-1	1	-1	-1106	-333	-257
344	1	4	1	-5	2	-2
8	d	o	1	5	-2	0
43	8	6	1	-43	8	38
	0	0	-1	-46	10	24
344	43	3	1	5719	-2150	946
344	n	o	1	-616	232	-102
43	8	6	1	2193	-688	-258
	0	0	-1	-70	42	-80
349	698	6	4	-65612	9074	7329
349	n	o	1	3548	-486	-387
349	-37	3	4	-2094	-2792	-1396
	0	2	-1	-912	-12	68
		c	-2	4886	698	0
				248	40	6
357	14	5	1	567	-301	392
357	d	o	2	-29	17	-20
7	-1	3	1	-367339	259182	-99246
	2	1	-1	15983	-5942	19262
360	1	6	1	7	-4	-5
40	d	o	2	1	0	-2
9	-3	3	1	-119	40	140
	-2	-1	0	-40	18	44
360	3	5	1	-21	-18	-15
120	d	o	2	3	4	2
9	-3	3	1	-22677	28260	-6840
	-2	-1	-1	-6360	2634	-2898
364	1	5	1	2	-1	-2
28	d	o	1	0	0	1
13	5	3	1	-629	-56	350
	-1	-2	-1	2	-122	-236

364	7	3	1	−14	−7	−7
28	d	o	1	−1	3	1
91	−16	6	3	28	14	49
	0	0	−1	57	−10	−1
364	7	5	1	−21	−21	−14
28	d	o	1	−34	−3	−1
91	11	9	3	−26994961	6817174	−3203382
	1	2	−1	10198242	−2574888	1214418
364	7	3	1	238	154	98
364	d	o	2	−25	−16	−10
7	−1	3	1	−175	−126	−28
	0	0	−1	18	16	10
364	13	5	1	−208	−169	−156
364	d	o	2	24	16	17
13	5	3	1	51519	−45136	38766
	−1	−2	−1	8094	−5862	−68
364	91	3	1	364	273	−91
364	d	o	2	−37	−29	9
91	−16	6	3	182	182	91
	0	0	−1	−61	−6	5
364	91	5	1	273	−273	182
364	d	o	2	−104	39	13
91	11	9	3	−84759493	11453806	35695842
	1	2	−1	−8873202	1195116	3742986
365	2	6	1	−8	1	−3
5	d	o	1	−6	1	−1
73	−7	9	1	−63	80	90
	−1	0	−1	109	−8	−30
365	146	6	1	511	73	146
365	d	o	2	−51	1	−8
73	−7	9	1	4287801	1473140	545310
	−1	0	−1	−228617	−76460	−29406
369	2	6	1	−14	7	−5
41	d	o	1	2	−1	1
9	−3	3	1	617	−369	−861
	−1	0	−1	123	−29	−115
371	2	6	3	−19	37	48
53	d	o	1	−1	3	8
7	−1	3	1	1857	−4452	−8162
	1	1	0	293	−620	−1098
			b1	−7	106	159
				−9	8	19
372	1	3	1	−11	−3	0
12	d	o	1	4	3	2
31	−4	6	1	157	69	27
	0	0	−1	−82	−43	−17

373	746	6	1	−5595	373	−1865
373	n	o	1	−103	9	55
373	−13	21	1	−148744194	14670836	26602360
	−1	0	−1	6643496	−933756	−1447124
377	2	6	1	66	−74	16
29	d	o	1	−18	10	−6
13	5	3	1	8354	−7076	2436
	−1	−1	0	−1692	1232	−508
377	26	5	1	273	208	169
377	d	o	2	−15	−10	−9
13	5	3	1	−14677	10933	−1131
	−1	−1	−1	−579	485	−337
380	19	5	1	589	−19	285
380	d	o	2	−43	12	−27
19	−7	3	1	−1750831	381900	−970330
	−1	1	−1	214722	−18878	104064
381	254	5	1	−4895977	1245743	−255524
381	n	o	1	251521	−63997	13126
127	20	6	1	−945007	−246126	−218694
	1	2	−1	53631	12334	9494
385	14	5	1	217	588	609
385	d	o	2	−11	−30	−31
7	−1	3	1	−1526	385	−2310
	1	2	−1	−104	81	−8
387	6	5	1	375	−261	57
129	n	o	1	−33	23	−5
9	−3	3	1	−1155	1161	−387
	−2	−1	−1	129	−69	57
387	258	5	1	99588	15867	15351
129	n	o	1	−8772	−1397	−1351
387	−39	3	3	150027	−69273	−20511
	1	−1	−1	40377	−1857	2223
387	258	5	1	299280	49020	17415
129	n	o	1	−26316	−4310	−1531
387	15	21	3	1248333	−135450	−222525
	2	1	−1	−112875	11940	19305
395	2	6	1	−21	−2	11
5	d	o	1	15	−2	−5
79	17	3	1	−2823	150	1180
	−1	0	0	1291	−46	−516
396	1	5	1	−5	−2	−1
44	d	o	1	1	1	0
9	−3	3	1	−8183	6072	−3366
	−1	−2	−1	−2508	1862	−950

397	794	4	1	9305283	-284649	1190603
397	n	o	1	-467897	14313	-59867
397	-34	12	4	-1191	-794	-397
	-2	0	-1	265	12	-11
399	2	3	1	7	3	6
21	d	o	1	1	-1	0
19	-7	3	1	11	24	60
	0	0	-1	15	-8	-4
399	14	3	1	-161	-14	-21
21	d	o	1	13	8	3
133	17	9	3	2079	-434	112
	0	0	-1	381	-94	40
399	14	5	1	175	35	7
21	d	o	1	53	3	-5
133	-10	12	3	-44485	4452	-9471
	-2	-1	-1	10791	-1222	1675
399	2	5	1	-10	-1	-4
57	d	o	1	0	1	0
7	-1	3	1	173	171	57
	1	2	0	-29	-19	-11
399	38	3	1	-152	19	0
57	d	o	1	-6	3	-2
133	17	9	3	-152	19	0
	0	0	0	-6	3	-2
399	38	5	1	-4142	38	646
57	d	o	1	-552	6	86
133	-10	12	3	-938410	101916	-170772
	-2	-1	0	124832	-13160	22856
403	2	6	1	-19	5	-6
13	d	o	1	-5	1	-2
31	-4	6	1	-180	-65	-481
	-1	0	-1	-230	83	33
403	26	6	1	1209	-247	-234
13	d	o	1	-283	67	72
403	-37	9	3	-9705490509	1467010090	1706398018
	1	0	-1	2670526035	-406406142	-476073822
403	26	6	1	-403	-52	-26
13	d	o	1	83	16	12
403	17	21	3	-28379	-4186	-2444
	0	1	-1	7251	1202	804
409	818	6	1	245507976	-9499843	29512213
409	n	o	1	-12139588	469737	-1459285
409	-31	15	1	-60532	-18814	2045
	1	1	-1	-7914	-744	-503

412	103	5	1	12171922	-1144227	2724865
412	n	o	1	-1199335	112744	-268489
103	-13	9	1	-2918397159	840663340	1194176850
	-1	1	-1	-287207118	82938090	117694800

413	14	5	1	2933	-1659	1974
413	n	o	1	-143	79	-102
7	-1	3	1	-199925439	449110242	809057914
	2	1	-1	-9829049	22091486	39817854

417	278	5	1	-255204	67554	-1529
417	n	o	1	12498	-3308	75
139	23	3	1	-573514	-62550	-93825
	1	-1	-1	-14606	-6600	-4275

420	1	5	1	5	-7	-8
60	d	o	2	0	-1	-3
7	-1	3	1	1	30	90
	0	3	0	-10	14	16

421	842	6	1	-74096	5052	-2947
421	n	o	1	3472	-238	129
421	-19	21	1	-471099	61466	117880
	1	1	-1	33591	-3642	-4624

423	6	5	1	-27	-30	-9
141	n	o	1	3	2	1
9	-3	3	1	-13530	10152	-5076
	-2	-1	-1	-1128	828	-468

427	2	6	1	-39	-30	-12
61	d	o	1	5	4	2
7	-1	3	1	37883	36234	12078
	0	1	-1	-5955	-4026	-2310

427	122	6	1	183	-61	61
61	d	o	1	-97	9	5
427	41	3	3	-2824483	41846	418460
	-1	0	-1	-323883	-266	53844

427	122	6	1	-36844	-5307	-488
61	d	o	1	-3798	-545	-52
427	-40	6	3	-35868	2257	-6039
	1	1	-1	7430	125	815

429	2	5	1	-39	-24	-17
33	d	o	1	7	4	3
13	5	3	1	-922	528	-297
	-2	-1	-1	-112	122	-29

429	26	5	1	2223	-1833	650
429	d	o	2	-113	85	-34
13	-5	3	1	-11253931	8943792	-3884166
	-2	-1	-1	572413	-444056	142850

```
433  866  6   1      74280038056      -6363464126        6019748726
433  n    o   1      -3569670822       305808570         -289290662
433  2   24   1   -275971151410      23641425888       -22426027796
     1    0  -1      13284676352      -1138109312        1073667456

437  38   5   1             -4294           665             -2242
437  n    o   1               210           -29              108
19  -7    3   1         -33327767        4883038         -17244894
    -2   -1  -1           1595651        -236062           822134

444   1   5   1                 7             3                 1
12    d   o   1                 4             1                 2
37   11   3   1              -713          -360              -390
     -1  -2   0              -584          -190              -110

444  37   5   1               703          -111              -407
444   d   o   2               -70            11                38
37   11   3   1          -3281789       -1279830          -1041624
     -1  -2  -1           -299554        -122716           -106838

453  302  3   1            -13137          3322              3775
453  n    o   1               487          -148              -203
151 -19   9   1            309701        -79728           -109324
      0   0  -1            -16407          3824              4788

455   2   6   1                71            22                 4
5     d   o   1               -33           -10                -2
91  -16   6   3               697           225                40
     -2  -1  -2              -343           -97               -22

455   2   6   1                73            24                18
5     d   o   1               -37           -10                -6
91   11   9   3            -56623        -19090            -14740
     -1   1  -2             30093          7894              4580

455   2   4   1                 3             5                15
65    d   o   2                 1            -1                -1
7    -1   3   1                 3             5                15
      0   0   0                 1            -1                -1

455  26   6   1            -19552          1066             -4888
65    d   o   2             -2274           174              -600
91  -16   6   3          15334046       -910260           4141800
     -2  -1   0           1976856       -138720            481368

455  26   6   1              -169           -52               -39
65    d   o   2               -11            -4                -3
91   11   9   3              -234            65                 0
     -1   1   0                22            -5                 6

456   1   5   1                 4            -1                 2
24    d   o   1                -2             0                -1
19   -7   3   1                13             0                12
     -1   1   0                -8             2                -2
```

456	19	5	1	−266	323	380
456	d	o	2	22	−30	−37
19	−7	3	1	−1861601	2088252	2645712
	−1	1	−1	185818	−188916	−246342
457	914	6	1	5967779226590	−413150621500	543860206786
457	n	o	1	−279160937624	19326370916	−25440707422
457	−10	24	1	369017446	−37574540	−70041648
	−1	0	−1	18986112	−1876952	−3119288
465	62	5	1	2046	−3658	1922
465	d	o	2	−124	186	−62
31	−4	6	1	445846320842	213958385400	84176161860
	−2	−1	−1	−20676308328	−9921712688	−3902991016
468	3	5	1	−57	3	15
12	d	o	1	27	0	−9
117	−21	3	3	−645	162	0
	−1	−2	0	−324	90	−18
468	3	3	1	57	−9	−9
12	d	o	1	12	−3	−9
117	6	12	3	57	−9	−9
	0	0	0	12	−3	−9
468	3	5	1	60	−45	21
156	d	o	2	−9	7	−4
9	−3	3	1	−16845	6786	18954
	−2	−1	−1	−2652	1026	3048
468	39	5	1	39	117	195
156	d	o	2	−117	−16	1
117	−21	3	3	−6789705	−1870362	−1748682
	−1	−2	−1	1032564	313920	278574
468	39	3	1	39	39	39
156	d	o	2	0	−7	−5
117	6	12	3	39	0	117
	0	0	−1	78	−12	−3
469	14	5	1	−7406	3724	−5026
469	n	o	3	342	−172	232
7	−1	3	1	483	938	3752
	2	1	−3	−89	106	96
469	134	5	3	−102561255	26127655	−18162494
469	n	o	3	4735839	−1206463	838666
67	5	9	1	−27487487	7022806	−5240606
	−1	1	−3	1318203	−334878	216222
			b2	17487	−5628	13601
				−2193	558	99
469	938	5	1	−427728	1876	−56749
469	n	o	3	19760	−88	2619
469	−43	3	3	−938	−1876	−3752
	−1	1	−3	1216	−92	−12

469	938	5	1	55580721	-1567867	-8850030
469	n	o	3	-2567301	72421	408788
469	38	12	3	1407	938	469
	-2	-1	-3	325	12	25
473	86	5	4	8856538	-3386680	1504140
473	n	o	3	-407224	155720	-69160
43	8	6	1	50998	-18920	9460
	0	0	-3	-2336	920	-360
		c	-3	-2752	1376	-430
				166	-48	30
476	7	5	1	-49	91	133
476	d	o	2	5	-8	-12
7	-1	3	1	-83055	-65926	-29988
	1	2	-1	-7514	-6088	-2650
477	2	4	1	20	7	15
53	d	o	1	0	-3	-1
9	-3	3	1	301	-240	210
	0	0	-1	57	-40	10
481	2	6	1	2	-2	-7
13	d	o	1	-4	0	1
37	11	3	1	220729	92690	83200
	-1	-1	-1	66143	25190	19800
481	26	6	1	-58097	-7722	-6682
13	d	o	1	-16031	-2142	-1868
481	41	9	3	72154200573	-9859570474	1277833908
	0	1	-1	-20879170581	2618986962	-454798716
481	26	6	1	1274	169	91
13	d	o	1	116	21	15
481	14	24	3	117	130	143
	-1	-1	-1	323	16	-9
481	2	6	1	4	-4	-5
37	d	o	1	0	0	1
13	5	3	1	-4586	2072	7252
	-1	-1	-1	788	-308	-1176
481	74	6	1	-16502	1332	4292
37	d	o	1	4852	66	-458
481	41	9	3	-15687033930	-2082293992	-1801577508
	0	1	-1	-2561317080	-342682548	-298886124
481	74	6	1	32782	4773	2849
37	d	o	1	-5396	-779	-461
481	14	24	3	71003	10730	6290
	-1	-1	-1	-12297	-1710	-1030
481	26	6	1	5239	1976	65
481	d	o	2	-239	-90	-3
13	5	3	1	-139945	77441	121693
	-1	-1	-1	1739	115	-6923

481	74	6	1	16909	−9694	1184
481	d	o	2	−771	442	−54
37	11	3	1	−4736	962	2405
	−1	−1	−1	−124	−8	117
481	962	6	1	−10944674	−1458873	−1267435
481	d	o	2	499028	66519	57791
481	41	9	3	−172030131	−23022103	−19937931
	0	1	−1	7875069	1045053	910623
481	962	6	1	−3999034	−578162	−344396
481	d	o	2	182368	26358	15696
481	14	24	3	−26197483182	1804109788	4442591036
	−1	−1	−1	−1195285424	82135208	202496232
483	2	5	1	73	−41	52
69	d	o	1	−9	5	−6
7	−1	3	1	−13591	−9936	−58374
	1	2	−1	4123	−4392	−3042
485	2	6	1	127	27	10
5	d	o	1	−31	−13	2
97	−19	3	1	15199757	−882730	4365080
	0	−1	−1	8063399	−33150	1987552
485	194	6	1	2231	291	485
485	d	o	2	31	−17	13
97	−19	3	1	2584274	549020	1934180
	0	−1	−1	287924	−36600	19612
488	1	6	1	−8	3	−2
8	d	o	1	9	−1	2
61	−1	9	1	−1767	476	−412
	−1	−1	−1	1494	−246	336
488	61	6	1	732	−61	122
488	d	o	2	−37	17	−6
61	−1	9	1	2941237	1140700	508740
	−1	−1	−1	288522	98550	51120
489	326	3	1	292633737	−5076472	61651164
489	n	o	1	−13233355	229566	−2787962
163	−25	3	4	−23146	3912	5379
	2	4	−1	−642	274	245
495	6	5	1	−33	21	30
165	d	o	2	3	−1	−2
9	−3	3	1	34161	40590	25740
	−2	−1	−1	2805	3054	2052
497	14	3	1	23408	18571	8428
497	n	o	1	−1050	−833	−378
7	−1	3	1	−3612	−3185	−1176
	0	0	−1	182	133	70

```
504    1   6   1         -11          -5           -1
  8    d   o   1          -9          -3            0
 63   15   3   3          73          36            0
      -1   1   0          72          24            6

504    1   4   1          11          -6            0
  8    d   o   1         -12           2           -2
 63  -12   6   3          11          -6            0
       0   0   0         -12           2           -2

504    3   5   1          27          -9          -15
 24    d   o   1          15          -5           -4
 63   15   3   3     4514331    -1587924     -1806948
      -1   1  -1     1845480     -647946      -737100

504    3   3   1          27           0           -6
 24    d   o   1          -6           2            4
 63  -12   6   3         255          90           72
       0   0  -1          84          42           30

504    1   5   1           8          -3           -7
 56    d   o   1          -2           1            2
  9   -3   3   1        -839        -756         -588
      -1  -2  -1        -168        -244         -134

504    7   5   1         -91          -7          -21
 56    d   o   1           7          -3            6
 63   15   3   3   -10152401      323484     -3984372
      -1   1  -1     3009720     -190618       946360

504    7   3   1          91         -28           14
 56    d   o   1         -28           8           -2
 63  -12   6   3        -581          42          252
       0   0  -1        -168          22           68

504    3   5   1          18           9          -21
168    d   o   2          -6           1            2
  9   -3   3   1     -230325      198828      -213948
      -1  -2  -1      -60312       40464        -4842

504   21   5   1        -315         147          147
168    d   o   2          63         -17          -22
 63   15   3   3     -109347      -44100        -5796
      -1   1  -1      -16968       -6726         -744

504   21   3   1        -315        -126          -84
168    d   o   2          42          20           16
 63  -12   6   3        1785         882          630
       0   0  -1        -336        -120          -96

508  127   3   1        4318        1905        -4699
508    n   o   1        -383        -169          417
127   20   6   1         254         254          127
       0   0  -1          91           6           13
```

```
511    2   6   1              18              -9             -25
 73    d   o   1              -2               1               3
  7   -1   3   1            -290             876            1825
       1   1  -1              66            -122            -191

511  146   6   1          134904           -2190          -20586
 73    d   o   1          -14958             362            2506
511   44   6   3     92654295014     -1328511816    -13809017844
       1   1  -1    -11136577032       118371432      1582282320

511  146   6   1            7081             803             292
 73    d   o   1            -821             -93             -34
511  -37  15   3          -34018            4015            5402
      -1   0  -1           -3880             515             620

515    2   6   1             -25               8              13
  5    d   o   1             -15               4               5
103  -13   9   1           -5263            -370             540
      -1  -1  -1             563             682             492

516    1   3   1             -40             -15              -9
 12    d   o   1              21               9               7
 43    8   6   1             586             294             171
       0   0  -1            -427            -136            -113

520    1   4   1              -7               4              -4
 40    d   o   2              -2               2               0
 13    5   3   1              -7               4              -4
       0   0   0              -2               2               0

520   13   4   1             429            -208            -572
520    d   o   4             -38              18              50
 13    5   3   1            -663            -364            -312
       0   0  -2             -54             -36             -22

527    2   6   1            -346             794             960
 17    d   o   1            -364              58             180
 31   -4   6   1   -36509568769030   -17520090317564   -6892400058996
      -1  -1  -1    8854891753360    4249241269608    1671660070216

531    6   5   1             318             -39            -240
177    n   o   1             -24               3              18
  9   -3   3   1          -20703            7965           23895
      -1   1  -1            1593            -645           -1785

532    1   3   1              -7               2               4
 28    d   o   1               0              -2              -2
 19   -7   3   1              -7               2               4
       0   0   0               0              -2              -2

532    7   5   1             287             -21              28
 28    d   o   1              26             -26              -1
133   17   9   3          113869          -27384           13426
       1   2   0           51684          -11070            2238
```

```
532    7  3  1              28               7                0
 28    d  o  1              11               3                2
133  -10 12  3              28               7                0
      0  0  0              11               3                2

532    1  5  1              -9               4               -9
 76    d  o  1               2              -1                2
  7   -1  3  1             761            -418             5.70
      -1  1 -1            -182             102             -116

532   19  5  1             152            -114             -209
 76    d  o  1            -187             -11               20
133   17  9  3       182911537        52532530         45021336
       1  2 -1        52055214        11207472          7011102

532   19  3  1            -589              57              -76
 76    d  o  1             120             -17               16
133  -10 12  3            5263            -323              893
       0  0 -1            -896             151             -179

533    2  6  1              39             -17              -66
 41    d  o  1              -7               3               10
 13    5  3  1         -466660          364203          1039924
      -1 -1 -1          123976          -25931          -139938

533   26  6  1             507             312              208
533    d  o  2             -21             -14              -10
 13    5  3  1          964223          589498           425334
      -1 -1 -1          -42117          -25270           -18466

536    1  6  1              -4               1               -1
  8    d  o  1               4              -1                0
 67    5  9  1             181             -48               52
      -1 -1 -1            -170              42              -16

536   67  5  1           11658           -2747             2010
536    n  o  1           -1006             237             -173
 67    5  9  1         1515607          538948           308736
      -2 -1 -1         -135932          -45286           -27510

541 1082  6  1           97380            1623             5951
541    n  o  1           -4078             -81             -251
541   29 21  1       -50654371         5151402         -2289512
       1  1 -1        -2150707          225538           -95312

545    2  6  1            -279              50              -44
  5    d  o  1             127             -22               20
109    2 12  1            -143              30              -20
       0 -1  0              69             -10               12

545  218  5  1          -87636           20710            -4796
545    d  o  2            4706            -618              348
109    2 12  1 -1360958690182422  -386123450046100  -203760496274260
       0 -1 -1  -58296246804552   -16539855481152    -8728011552424
```

549	2	6	1	16	-27	2
61	d	o	1	-4	1	-2
9	-3	3	1	-1645	122	-854
	-1	-1	0	61	-186	-2
549	122	6	1	148718	20496	3355
61	d	o	1	-17690	-2438	-399
549	42	12	3	122	122	61
	-1	0	0	-122	-4	5
549	122	6	1	-18727	-2867	-2196
61	d	o	1	2623	343	288
549	-39	15	3	6954671	-780434	198860
	1	1	0	898591	-99018	26020
551	2	6	1	103	-10	63
29	d	o	1	-21	4	-9
19	-7	3	1	-2717675	2890894	3706896
	-1	-1	-1	-500429	536146	690568
553	14	5	1	102459	-188692	-283038
553	n	o	1	-4357	8024	12036
7	-1	3	1	-6622	-5530	-2765
	1	2	-1	-286	-220	-85
553	158	5	1	-45868664	16010930	-1928943
553	n	o	1	1950534	-680854	82027
79	17	3	1	465784	-185255	53088
	-1	1	-1	-26120	8151	310
553	1106	5	1	-3222884	655858	2260664
553	n	o	1	137090	-27894	-96140
553	-22	24	3	-3134859900942	-382865477708	-163457463148
	-1	1	-1	113052877472	17321642392	5086836360
553	1106	5	1	-46492074251	3461407278	7376408935
553	n	o	1	1977044101	-147194010	-313676815
553	5	27	3	-89751900	-11993464	-6196365
	-2	-1	-1	-3697422	-518886	-282603
556	139	5	7	1610454	-420614	35723
556	n	r2	1	-136597	35676	-3030
139	23	3	1	139	-278	278
	2	1	-1	-170	26	20
559	2	6	1	-21	8	-3
13	d	o	1	5	-2	1
43	8	6	1	-11	0	13
	-1	0	0	-7	2	3
559	26	6	1	-1703	-208	-195
13	d	o	1	-479	-56	-55
559	47	3	3	23855	3666	3094
	1	0	0	8449	782	866

559	26	6	1	12168	-754	-1638
13	d	o	1	2642	-310	-500
559	-7	27	3	-4734626	408044	765648
	-1	-1	0	-1351488	115764	209316
572	1	5	1	8	43	8
44	d	o	1	19	0	7
13	5	3	1	-10371371	8030682	-3082486
	-2	-1	-1	3079046	-2450452	908250
572	13	5	1	-182	-13	104
572	d	o	2	1	-8	-15
13	5	3	1	-22058023	-13399958	-9779198
	-2	-1	-1	-1851178	-1117812	-807526
577	1154	6	1	-1629923448	133393168	-102192470
577	n	o	7	67854598	-5553230	4254328
577	11	27	1	-33855231506	-4527070452	-2566694488
	-1	0	-7	1409115276	188490216	106812948
581	14	3	1	10080	8071	3591
581	n	o	1	-418	-335	-149
7	-1	3	1	10143	1330	5292
	0	0	-1	-5	-286	68
584	1	6	1	-7	2	0
8	d	o	1	-7	1	0
73	-7	9	1	717	280	84
	-1	-1	0	-614	-182	-84
584	73	5	1	-2263	730	1168
584	d	o	2	191	-61	-96
73	-7	9	1	10877	-3796	-6424
	-1	-1	-1	-1092	322	490
585	2	6	1	-59	1	18
5	d	o	1	-29	1	8
117	-21	3	3	53627	-1500	-15750
	-1	-2	-2	24375	-580	-6950
585	2	6	1	-59	10	18
5	d	o	1	-29	4	8
117	6	12	3	-26518	4890	8145
	1	2	-2	-11310	2108	3745
585	2	6	1	16	-7	-16
65	d	o	2	-2	1	2
9	-3	3	1	-2533	-2145	-975
	-1	1	-2	-195	-313	-257
585	26	6	1	143	78	104
65	d	o	2	-13	-10	-14
117	-21	3	3	-1158469	293085	-35295
	-1	-2	-2	-138255	37943	-2915

```
585   26   6   1          7462            1560              286
 65    d   o   2           520             268              158
117    6  12   3    -482335727614    64399329540     -89412264240
       1   2  -2     59835154080     -7984949992      11091403744

589   38   5   4         -22933           24149            30628
589    n   o   1           945            -995            -1262
 19   -7   3   1          -5263           -3534           -1178
       0   0  -1          -239            -118             -10
            c2            -171            -209            -114
                          -13              -3               2

589   62   5   1          -8959           2015            -1798
589    n   o   1           369            -83               74
 31   -4   6   1          1240            4123            -589
      -2  -1  -1          -458            -75             -121

589 1178   5   1         7071534        -723881           67735
589    n   o   1         -291378          29827           -2791
589   41  15   3       -471416163      -25428308       -36872578
      -2  -1  -1         -7766645        -2342740        -1199190

589 1178   5   1        -40281121       -7932063        -4686673
589    n   o   1         1659755         326835          193111
589  -13  27   3       1105991805      766027484       6745872366
      -1   1  -1       2376826365      -175641564       -80227362

595    2   6   1          -14              9              -13
 85    d   o   2           -2              1               -1
  7   -1   3   1          -253            170             -170
       1   2   0           -31             14              -22

597  398   5   1        2929479        -308251         -802368
597    n   o   1         -119903         12617           32838
199   11  15   1       -166103509      -33394986       -19049076
       2   1  -1         -6199055        -1429782        -943908

601 1202   6   1     21825163864126  -2040868029070   1036234153684
601    n   o   1     -890267003356    83248743322     -42268872780
601   26  24   1     11784034467682  -1102014723948    559479787588
       0  -1  -1     -480714546664    44947852760     -22824429888

603    6   5   1          -1659           312            -822
201    n   o   1           117            -22               58
  9   -3   3   1          -37983         -16281           4824
      -1   1  -1           201            2127            2478

603  402   5   1        1799621742      -18178842      -228586446
201    n   o   1        -126935520       1282238        16123246
603  -48   6   3    -44143909920390    442230489288    5600525018148
       1  -1  -1      3113662615584    -31190426832    -395030187120

603  402   5   1          -402           -14271         -10653
201    n   o   1            0             1009            755
603   33  21   3         121605         -12663          -18693
      -2  -1  -1          9849           -915            -1203
```

604	151	5	1	9701146	-575914	1953789
604	n	o	1	-789459	46865	-159000
151	-19	9	1	-837020331	201886094	260485268
	2	1	-1	-65646954	16278288	21683094
609	14	3	1	-469	1407	-567
609	d	o	2	19	-57	23
7	-1	3	1	-49	-147	-189
	0	0	-1	5	-1	-5
612	3	3	1	-18	18	-12
204	d	o	2	3	-2	2
9	-3	3	1	-69	72	-18
	0	0	-1	12	-6	6
613	1226	6	1	-943407	-449329	-226810
613	n	o	1	153941	4613	10946
613	47	9	1	-593101722695	-136560476712	295481201174
	-1	-1	-1	140376795117	-8087885472	-10140130314
616	1	5	3	10	-13	-17
88	d	o	1	0	-2	-5
7	-1	3	1	-351	748	1276
	1	2	0	-62	148	282
			b2	-111	-88	-44
				24	18	8
616	7	5	1	224	371	35
616	d	o	2	-18	-30	-3
7	-1	3	1	-108101	-88704	-41580
	1	2	-1	8814	6906	2904
620	31	3	1	1364	-310	465
620	d	o	2	-107	26	-37
31	-4	6	1	3286	-775	930
	0	0	-1	-223	51	-100
623	2	6	1	-9	18	-11
89	d	o	1	1	-2	1
7	-1	3	1	269	1691	2225
	0	1	-1	-131	51	179
627	2	3	1	-16	12	15
33	d	o	1	2	-2	-3
19	-7	3	1	83	-69	-87
	0	0	-1	-9	13	17
629	2	6	1	16	-5	-1
17	d	o	1	-2	1	-1
37	11	3	1	3827	-5049	-11781
	0	-1	-1	5397	561	-1353
629	74	6	1	5069	-444	-3108
629	d	o	2	-183	24	130
37	11	3	1	-4520300249	-1807065422	-1523001474
	0	-1	-1	-180336349	-72042646	-60657970

```
632    1  6  3          -70          36              1
  8    d  r  1          -75          18             -6
 79   17  3  1         -107          36              0
       0  1  0          -66          24             -6

632   79  5  1        35471       10191           9085
632    n  o  1        -2804        -817           -722
 79   17  3  1    -66008529    23300892       -3191284
       1  2 -1     -5389358     1859894        -197352

633  422  5  1    -91413640    15698189       24736585
633    n  o  1      3633366     -623947        -983191
211  -13 15  1      -467998     -134829         -68364
      -1  1 -1       -25880       -4107           -738

635    2  4  1           15          -3             -1
  5    d  o  1           -3           1             -1
127   20  6  1          149          -4            -51
       0  0 -1           75          -6            -23

639    6  5  4         -846         639           -426
213    n  o  1           66         -47             20
  9   -3  3  1       -53031       20448          60066
       0  0 -1        -3621        1464           4134
             c3          96         -63           -177
                         12          -3             -9

644    1  5  1           32          39              6
 92    d  o  1          -11          -6             -5
  7   -1  3  1       -19457      -14306          -7268
      -1  1  0         3658        3204           1238

645   86  5  1       -13803        2838           9116
645    d  o  2          545        -112           -360
 43    8  6  1        24596       10965           7095
      -1 -2 -1         1086         389            289

651    2  5  1          -74          41             67
 21    d  o  1           16          -9            -15
 31   -4  6  1         1115        -672          -1197
      -1 -2 -1         -301         158            241

651   14  5  1           42          -7             -7
 21    d  o  1          -22           1              3
217   29  3  3      -557977      116592          -8778
      -1 -2 -1       122037      -25528           1690

651   14  5  4          791        -231           -294
 21    d  o  1          281         -43            -52
217  -25  9  3      -115633         378          12684
       0  0 -1         -277        5170           3764
             c2          476          70             21
                         -80         -18             -5

651    2  5  1          -54         -20            -26
 93    d  o  1           -2          -4              0
  7   -1  3  1        -3346       -2976          -1116
       2  1  0         -392        -284           -148
```

```
651   62  5  1        -1395         -31         -124
 93    d  o  1           -5          27           20
217   29  3  3       447113      236406       138942
      -1 -2  0      -127183       -7618       -15602

651   62  5  4         -961           0         -186
 93    d  o  1           95          -8           14
217  -25  9  3         -961           0         -186
       0  0  0           95          -8           14
             c3         -527          93           93
                        -25           7           11

652  163  5  4      -574412       13040      -115078
652    n  o  1        44990       -1021         9014
163  -25  3  4          489         652          326
      -2  2 -1         -212          -6           22
             c2          815         163            0
                        -60         -14           -3

657    2  4  1           25          13           17
 73    d  o  1           -1          -3           -1
  9   -3  3  1           25          13           17
       0  0  0           -1          -3           -1

657  146  6  1          146           0            0
 73    d  o  1         -146         -18           -2
657   51  3  9          146        -292         -584
       1  2  0          584         -36           -4

657  146  6  1      -255938       12118        35624
 73    d  o  1       -27448        1750         4396
657  -30 24  9     10615514    -1646588      2917664
      -1 -2  0     -4566880      332008        98080

661 1322  6  1    -12705081      122946     -1451556
661    n  o  1       488339       -4704        55842
661  -49  9  1   -193632679   -71982900    -33112134
      -1  0 -1    -23756139     -880044       874698

665    2  6  1          216          54           40
  5    d  o  1           98          24           18
133   17  9  3         -218         -60          -40
       1  2  0         -108         -24          -20

665    2  4  1            9           3            2
  5    d  o  1           -5          -1            0
133  -10 12  3            9           3            2
       0  0  0           -5          -1            0

665   14  5  1         -672         210         -406
665    d  o  2           26          -8           16
  7   -1  3  1       -18606       31920        61180
      -1  1 -1         -472        1444         2460

665   38  3  1         7182       -7505        -9690
665    d  o  2         -278         291          376
 19   -7  3  1         7258       -5130        -7505
       0  0 -1         -156         268          309
```

665	266	5	1	−7182	−1463	−665
665	d	o	2	296	61	29
133	17	9	3	253631	75145	65835
	1	2	−1	12731	2669	1609
665	266	3	1	261478	−59584	46284
665	d	o	2	−8928	2040	−2224
133	−10	12	3	78259062	−33942132	−43032416
	0	0	−1	5668784	−627272	−1414104
669	446	5	1	2307381	−577570	−753963
669	n	o	1	−88509	22310	29269
223	−28	6	1	110831	2007	30774
	−2	−1	−1	6355	−355	688
673	1346	6	1	−15784891960	529192687	−1464703740
673	n	o	1	608462870	−20398879	56460180
673	−37	21	1	−96321779	2944375	−8833125
	−1	0	−1	3573575	−130725	335825
679	2	4	1	89	−49	69
97	d	o	1	−9	5	−7
7	−1	3	1	29	−23	11
	0	0	−1	−3	1	−3
679	194	6	1	10864	194	679
97	d	o	1	−1028	−12	−65
679	23	27	3	−2363459805	119500799	340324209
	0	−1	−1	−250727691	13038585	34029921
679	194	4	1	931976	128040	67512
97	d	o	1	106834	12128	5152
679	−4	30	12	499162	−121832	39188
	−2	−2	−1	−148040	224	−9784
684	3	3	1	−150	12	−15
12	d	o	1	45	2	19
171	24	6	3	8211	−1755	−2070
	0	0	−1	4902	−1005	−1170
684	3	5	1	321	84	45
12	d	o	1	183	49	26
171	−3	15	3	86187	−19116	10350
	−2	−1	−1	−69540	5802	−8766
684	1	3	1	−25	8	22
76	d	o	1	−4	2	6
9	−3	3	1	−25	8	22
	0	0	0	−4	2	6
684	19	3	1	19	38	19
76	d	o	1	−38	−2	3
171	24	6	3	19	38	19
	0	0	0	−38	−2	3

```
684   19  5   1          -19              -95              -76
 76    d  o   1           95               14               -3
171   -3 15   3       225283           -27550           -66386
      -2 -1   0        59660            -7438           -14244

689    2  4   1           25               14                6
 53    d  o   1           -3               -2               -2
 13    5  3   1           25               14                6
       0  0   0           -3               -2               -2

689   26  3   1         7904            -3263           -12324
689    d  o   4         -300              125              470
 13    5  3   1        44642            27664            20163
       0  0  -2         1740             1030              745

693    6  3   1           -9                3               -9
 33    d  o   1            3               -1                1
 63   15  3   3          -57               45               36
       0  0  -1          -21                3                6

693    6  5   1         2718             -750              360
 33    d  o   1         -444              158              -32
 63  -12  6   3  36911992687422   -3783173935860   -15489565496460
      -1  1  -1   6447461220816    -665486607240    -2694133902240

693    2  5   1         -177               48              189
 77    d  o   1          -19               10               23
  9   -3  3   1    -93207805       -121417758        -88103862
      -1 -2  -1     12916827         12930546          7430802

693   14  3   1         -210               -7              -56
 77    d  o   1           14               -3                6
 63   15  3   3         3101             -126             1008
       0  0  -1         -315               18             -120

693   14  5   1        -1974             -826             -623
 77    d  o   1          224               94               71
 63  -12  6   3       -19390             3465            -2772
      -1  1  -1        -1386              741              -54

695    2  6   3          -76               13               11
  5    d  o   1            0               -5                5
139   23  3   1       -57623             1250            17000
       1  0   0       -26195              690             7600
             b2         -304              -25               25
                        -22               15               35

703    2  4   1          100               20              -28
 37    d  o   1           10               12                8
 19   -7  3   1        10102             6652              860
       0  0  -1         2028             1040              332

703   74  6   1      -255929           -28786            -2109
 37    d  o   1        42355             4764              349
703  -52  6   3       -27047            -2960             -370
      -1  0  -1         4147              480               10
```

```
703    74   4   1        33633           3996          1443
 37     d   o   1        -5585           -684          -235
703   -25  27  12      1807450        -159988       -243904
       -2   0  -1      -287076          25800         40932


707     2   6   1           13             15            84
101     d   o   1            5             -5            -4
  7    -1   3   1     -10258770       23171824      41875812
        1   1  -1      1035008        -2313644      -4156940


709  1418   6   4     -11241904       -1197501      -1152125
709     n   o   1       422200          44973         43269
709    53   3   4        -1418           2836         -2836
        2   2  -1        -1820            112           100
                c-3      12762          -1418             0
                          458            -50             6


711     6   5   1        -4962           3705         -1956
237     n   o   1          324           -239           128
  9    -3   3   1   -4167474591     3090729798   -1634904684
       -2  -1  -1    269777337     -200397378     107255292


711   474   5   1      -5785881         186993        724983
237     n   o   1       376593         -12055        -47027
711   -39  21   3    4696459071      597166056     439115022
        1   2  -1    310844697       38253360      28709226


711   474   3   1     15278442        -800823      -1914960
237     n   o   1       758400         -46469       -102388
711   -12  30  12          474            711           711
       -4  -2  -1         -474            -21             9


713    62   3   1     37320838       -8683720      13391132
713     n   o   1     -1397648         325192       -501528
 31    -4   6   1   1310333234     -733140452   -1208515656
        0   0  -1     49046536      -27456816     -45266984


721    14   5   1      3470012       -6287883     -11271484
721     n   o   1      -129230         234173        419772
  7    -1   3   1          735          -5047         -7931
        1   2  -1         -111            123           265


721   206   5   1    19180802861   -1802622982    4293573231
721     n   o   1     -714330427      67133188    -159901023
103   -13   9   1     -538531277     156883111     221855305
       -2  -1  -1      -20256471       5784087       8245203


721  1442   5   1    -2424205322    -280048657    -234184405
721     n   o   1      90282124       10429557       8721483
721    47  15   3      53896913       -1310057      -6991537
       -1   1  -1       2057715         -43451        -255821


721  1442   5   1  -1835163664880  -218703929360  -65299693830
721     n   o   1    68345066518     8144959976    2431887686
721   -34  24   3    130718742       10454500      69792800
       -1   1  -1     -24485560        1472360         54960
```

```
728    1   4   1        -93          -2          -20
  8    d   o   1        -36           8          -12
 91  -16   6   3        -93          -2          -20
      0   0   0        -36           8          -12

728    1   6   1          5           0           -1
  8    d   o   1         -4           0            1
 91   11   9   3         33         -12            4
      1   2   0         34          -6            4

728    1   5   1        -44         -41          -47
 56    d   o   1        -19          -8           -1
 13    5   3   1     -15539      -19432       -26124
     -1  -2   0      -9372       -2998         1034

728    7   3   1         77           0           14
 56    d   o   1        -10           2           -4
 91  -16   6   3         77           0           14
      0   0   0        -10           2           -4

728    7   5   1       -119          14           49
 56    d   o   1        -34           4           13
 91   11   9   3        903        -252          196
      1   2   0       -334          78          -16

728    1   4   1          5           4           12
104    d   o   2          0           0           -2
  7   -1   3   1        -51         -96           -4
      0   0  -2        -26         -10          -12

728   13   4   1       1443        -494         -624
104    d   o   2       -282          98          122
 91  -16   6   3       3341        3302         1950
      0   0  -2       2046         168         -218

728   13   6   1       -143           0           39
104    d   o   2        -16           6           11
 91   11   9   3    -285831       44720       129324
      1   2  -2     -61624        7106        24238

728    7   3   1        217         392           84
728    d   o   2        -36         -18          -20
  7   -1   3   1     -24381       11340       -22960
      0   0  -1      -1996        1270         -930

728   13   5   1      -1612        1261         -481
728    d   o   2        119         -94           35
 13    5   3   1     -81887     -177268      1190280
     -1  -2  -1     -92214       54484        62722

728   91   3   1     -15379       -4732         -910
728    d   o   2       1172         348           74
 91  -16   6   3    -201747        5278       -47684
      0   0  -1     -11826        1346        -3336
```

728	91	5	1	5005	−546	−1911
728	d	o	2	−326	54	151
91	11	9	3	3269175	1225952	697788
	1	2	−1	311620	73378	59954
731	2	6	1	3712	−850	−2272
17	d	o	1	782	−138	−568
43	8	6	1	−476276826	97357164	316702792
	−1	0	0	−115454216	23637248	76828376
732	1	5	4	17	−18	−24
12	d	o	1	−32	2	10
61	−1	9	1	17	−18	−24
	0	0	0	−32	2	10
			c−3	−15	−4	−1
				−6	−3	−2
732	61	5	4	−61	−366	0
732	d	o	2	8	26	−2
61	−1	9	1	−671	−1098	−1098
	0	0	−1	242	24	−30
			c2	671	−183	−366
				47	−15	−27
733	1466	6	1	−2226310114	232476081	−38386477
733	n	o	3	82220864	−8585655	1417677
733	50	12	1	2199	1466	733
	0	1	−3	503	12	31
737	134	5	1	6776782	−1725049	1199903
737	n	o	1	−249626	63543	−44199
67	5	9	1	16348	−4422	3685
	−1	1	−1	−712	172	−97
741	2	5	1	−76	60	−23
57	d	o	1	10	−8	3
13	5	3	1	1142	−969	684
	−1	−2	−1	−202	149	−14
741	38	5	1	2394	−399	−475
57	d	o	1	−316	53	63
247	−31	3	3	584630	−125115	−134406
	−1	1	−1	−79294	16553	17512
741	38	5	4	572014	−58140	67260
57	d	o	1	−75856	7680	−8920
247	−4	18	3	−27741862	2724600	−4292100
	0	0	−1	4420360	−460280	383360
			c2	−1387380	140752	−163134
				183770	−18640	21610
741	26	5	1	6916	−3003	−10816
741	d	o	2	−262	107	394
13	5	3	1	−91019200363	−54973914138	−39733999812
	−1	−2	−1	−3332741953	−2024115822	−1476455172
741	38	5	1	3876	−551	2185
741	d	o	2	−148	17	−81
19	−7	3	1	−75959131	6789042	−36870678
	−2	−1	−1	2345503	−506754	1297374

```
741   494  5   1        -39273           -741           3458
741    d   o   2          1449             27           -126
247   -31  3   3      -18786079      -12604410      -28930122
      -1   1  -1       -5404365         533322           -558

741   494  5   4           247           -741              0
741    d   o   2             7             25             -4
247    -4 18   3          -988          -2223          -2223
       0   0  -1           490             21            -33
               c2        -2470           -247           -247
                           668            -49             77

744    1   5   4            41             24             12
 24    d   o   1           -20             -8             -2
 31   -4   6   1         -1595           -756           -312
       0   0  -1           638            310            116
               c4         -127            -62            -24
                            53             25             10

744   31   3   1          1643           -186           -744
744    d   o   2          -122             14             54
 31   -4   6   1         -2945           1302           1302
       0   0  -1            54            -56           -154

747    6   5   1          2793           2793           1941
249    n   o   1          -177           -177           -123
  9   -3   3   1         24657          38097          20169
      -2  -1  -1         -2241          -1911          -1545

749   14   5   1          1127          -2597          -4760
749    n   o   1           -41             95            174
  7   -1   3   1      -7743897       -6240668       -2808750
       1   2  -1       -283673        -226372         -99598

755    2   6   3           202            -46            -62
  5    d   r   1           -84             22             28
151  -19   9   1           -58            -20              0
       1   1   0           -44             -8             -4

757 1514   6   1       3065850         177895         205147
757    n   o   1        -46898         -11825          -4721
757   29  27   1  -4383949447658   376429668588  -241013775064
      -1   0  -1  -182481935496    14705970204    -5728198428

760    1   6   1            -6              1              0
 40    d   o   2             0              0             -1
 19   -7   3   1        -13899          14740          18800
      -1   1  -2          4310          -4632          -5998

760   19   5   1         -2964           3059           3990
760    d   o   2           216           -222           -289
 19   -7   3   1      -35606361       40506100       51123680
      -1   1  -1       2803078       -2811396       -3680690

763    2   6   1            18            -20            -32
109    d   o   1             0             -2             -4
  7   -1   3   1           874          -2180          -3924
       1   1   0           100           -204           -368
```

763	218	4	1	5886	−109	−436
109	d	o	1	526	−21	−36
763	−55	3	12	−12099	436	−436
	−2	0	0	453	36	124
763	218	6	1	−2770453	−294736	−263889
109	d	o	1	265269	28236	25281
763	53	9	3	−128990491	−14238234	−12490092
	1	1	0	12818013	1314990	1202076
765	2	4	3	−107	60	150
85	d	o	2	−15	4	14
9	−3	3	1	−107	60	150
	0	0	0	−15	4	14
			b2	14	20	15
				−2	−2	−1
769	1538	6	1	−1376850667	21422802	−141868196
769	n	o	1	49650505	−772526	5115898
769	−49	15	1	−25481584	1787156	−3133675
	0	1	−1	−1623696	−14334	−126635
776	1	6	1	−33	−7	1
8	d	o	1	19	7	2
97	−19	3	1	−4059	−1912	−612
	0	−1	0	4516	758	−226
776	97	5	1	−105827	3007	−28033
776	d	o	2	7581	−221	2012
97	−19	3	1	3079813341	−81524620	830809656
	0	−1	−1	−224214778	6970168	−58410206
777	2	5	1	38	27	16
21	d	o	1	14	3	4
37	11	3	1	653	378	252
	−2	−1	0	199	54	60
777	14	5	1	−31710	−6020	−1848
21	d	o	1	6808	1332	430
259	−19	15	3	−1526938	173292	300300
	−1	−2	0	−239604	55856	71212
777	14	3	1	77	−7	0
21	d	o	1	−5	1	−2
259	8	18	3	77	−7	0
	0	0	0	−5	1	−2
777	14	5	1	2359	1288	−35
777	d	o	4	−85	−46	1
7	−1	3	1	−5056702	2707845	−3774666
	1	2	−2	−185568	106513	−118540
777	74	5	1	−4810	222	2923
777	d	o	4	172	−8	−105
37	11	3	1	−307618	−693084	−1236207
	−2	−1	−2	−92118	−16358	9623

777	518	5	1	-2662779	-513338	-161616
777	d	o	4	95527	18416	5798
259	-19	15	3	808598	162393	48174
	-1	-2	-2	-30762	-5693	-1900
777	518	3	1	13632206	-1685572	1258740
777	d	o	4	-488136	60376	-45376
259	8	18	3	2036910162	-206957576	-485400188
	0	0	-2	72835152	-7469064	-17439896
779	2	6	1	-4	5	11
41	d	o	1	2	-1	-1
19	-7	3	1	-24352	26199	33948
	-1	0	-1	3888	-4119	-5250
780	1	3	1	-19	18	-6
60	d	o	2	6	-4	2
13	5	3	1	-173	144	-42
	0	0	-2	46	-34	16
780	13	3	1	1417	-624	-2262
780	d	o	4	-106	42	160
13	5	3	1	-35009	-21138	-14976
	0	0	-2	-2492	-1514	-1126
785	2	6	1	-395	75	-36
5	d	o	1	-175	33	-16
157	14	12	1	-23998	2335	7305
	-1	-1	-1	10572	-1099	-3295
785	314	6	1	-604450	75360	161082
785	d	o	6	23024	-2340	-5518
157	14	12	1	47790458054	-7273345280	-18142759860
	-1	-1	-3	-2229391056	136281896	564505016
791	2	4	1	63	57	17
113	d	o	1	-7	-5	-3
7	-1	3	1	63	57	17
	0	0	0	-7	-5	-3
792	1	5	1	-16	9	21
88	d	o	1	4	-1	-4
9	-3	3	1	-263	220	440
	-1	-2	0	88	-10	-70
792	3	5	1	-114	93	-69
264	d	o	2	18	-13	4
9	-3	3	1	8099787	-3204036	-9219276
	-1	-2	-1	998184	-393576	-1134042
793	2	6	1	10142	-1993	2304
13	d	o	1	-2658	609	-612
61	-1	9	1	7060133	-1606878	1624428
	-1	-1	0	-2061771	407970	-468684

793	26	6	1	3003	403	130
13	d	o	1	−1007	−95	−14
793	−43	21	3	−273	5018	3718
	1	0	0	−13361	−102	670
793	26	6	1	6695	754	546
13	d	o	1	1503	166	118
793	38	24	3	−3029	130	338
	1	1	0	685	−18	−98
793	2	6	3	35	72	113
61	d	o	1	17	4	−5
13	5	3	1	20315	29402	41846
	0	−1	0	6867	1970	−1194
			b2	−266	−183	−61
				−36	−19	−23
793	122	6	3	4392	488	122
61	d	r	1	−554	−62	−14
793	−43	21	3	−1159	0	−122
	1	0	0	−143	8	−10
793	122	6	3	−98881	−11041	−7930
61	d	o	1	12425	1387	996
793	38	24	3	305	427	122
	1	1	0	−543	−11	−32
			b2	−61	61	61
				−25	−5	−9
793	26	5	1	338	−9659	−676
793	d	o	4	−12	343	24
13	5	3	1	−7111	−5551	−5551
	0	−1	−2	339	161	63
793	122	5	1	−92659	43371	70760
793	d	o	4	3285	−1539	−2514
61	−1	9	1	−12275525686	3638656710	7013494215
	−1	−1	−2	435454872	−129111960	−249162315
793	1586	5	1	4949906	287066	−64233
793	d	o	4	−175776	−10194	2281
793	−43	21	3	49438792	−4996693	−7048184
	1	0	−2	−2016674	184693	227442
793	1586	5	1	14274	166530	39650
793	d	o	4	816	−5766	−1302
793	38	24	3	−1773105515818	165724337376	−95025186828
	1	1	−2	78235015704	−6425839344	1455124728
796	199	5	1	−87211750	13220565	−7429665
796	n	o	1	6182277	−937181	526675
199	11	15	1	5066540199	−774716950	482425750
	2	1	−1	−359281210	54917570	−34181700
801	2	4	1	−96	−112	−76
89	d	o	1	10	12	8
9	−3	3	1	238	−156	108
	0	0	−1	24	−20	8

804	1	3	1	-10	2	2
12	d	o	1	1	0	-2
67	5	9	1	147	-30	-68
	0	0	-1	-70	16	42
805	14	5	1	-3507	-2800	-1211
805	d	o	2	123	100	45
7	-1	3	1	-264779781	-212643970	-94420060
	-1	1	-1	9348263	7485902	3339132
812	7	3	1	329	378	196
812	d	o	2	-26	-20	-2
7	-1	3	1	-20657	-18494	-10178
	0	0	-1	1538	1098	352
813	542	5	4	-247423	-43902	-35772
813	n	o	1	8615	1542	1268
271	29	9	1	2637914	471540	390240
	0	0	-1	-94492	-16360	-13460
			c2	235228	37940	33062
				-7546	-1450	-1134
815	2	4	1	-71	-5	-10
5	d	o	1	9	-3	4
163	-25	3	4	3203	-40	690
	0	0	-1	-1447	32	-298
817	38	5	1	4023573	2646491	845177
817	n	o	5	-140767	-92589	-29569
19	-7	3	1	-204212	72713	-17974
	-1	1	-5	-3904	-917	-5076
817	86	5	1	38781184	-50897896	1660058
817	n	o	5	-1356782	1780692	-58078
43	8	6	1	-4021477047000094	-1606462649791196	-1112415479834344
	-2	-1	-5	140693356226584	56203078385152	38918702008696
817	1634	5	1	-47299132827127	-5004603447455	-506514161558
817	n	o	5	1654787189961	175088911375	17720687378
817	-55	9	3	-4879704070	50961192	-473230093
	2	1	-5	-172071534	1927686	-16395627
817	1634	5	1	-9458659002	617718994	1221849644
817	n	o	5	330916590	-21611252	-42747108
817	-1	33	3	-622356286	-70428668	-34474132
	-2	-1	-5	-21500620	-2481784	-1241812
819	2	6	1	19	11	1
13	d	o	1	7	3	1
63	15	3	3	-1909	182	-728
	-1	1	0	637	-6	208
819	2	6	1	275	116	88
13	d	o	1	77	32	24
63	-12	6	3	275	130	104
	-1	1	0	91	34	24

```
819   26  6  1                26            -39            -39
 13    d  o  1               182             -1              3
819  -57  3  9           -223561         -38662         -51194
      -1  1  0            136591          10450           6246

819   26  6  1             28418           -546           2652
 13    d  o  1              7826           -146            742
819   51 15  9           -227110           3952         -21268
      -2 -1  0            -61880           1268          -5848

819   26  6  1             92248          -4121           7007
 13    d  o  1             26572          -1195           1967
819   24 30  9                26            -13            -65
      -2 -1  0              -182             11              5

819   26  6  1             28366          -1768           1651
 13    d  o  1             -7826            488           -463
819   -3 33  9           -138385           8398           3224
      -2 -1  0              3367           -274           3344

819    6  3  1               282             -9            -75
 21    d  o  1               -54              1             17
117  -21  3  3              4569            234           -702
       0  0 -1              -507             90            282

819    6  5  1               186             45             12
 21    d  o  1               -24            -13             -8
117    6 12  3                 6            567            -63
       1  2 -1              -546            -51            -87

819   42  5  1             16716          -1722             63
 21    d  o  1             -3654            376            -13
819  -57  3  9          -5864817           9324         638694
      -1  1 -1          -1333059           7548         139194

819   42  5  1           -108444           2037         -10206
 21    d  o  1            -23646            451          -2224
819   51 15  9     1699800061911    -32180435280   159907079178
      -2 -1 -1      370934274165     -7022985960    34893780630

819   42  5  1               483            -63           -231
 21    d  o  1             -7539            329           -623
819   24 30  9         -11557686        -1391796        -516411
      -2 -1 -1           2541630         302862         114303

819   42  5  1              -483             21             21
 21    d  o  1              -105              5            -17
819   -3 33  9       -1058443617      -128247966      -66141810
      -2 -1 -1        -231330099       -27964962      -14389206

819    6  5  1              -582            222            666
273    d  o  2                36            -14            -40
  9   -3  3  1          -2555274        1015560        2912364
      -2 -1 -1            155064         -60852        -175932
```

```
819   42   5   1              -399                    42                   -21
273   d    o   2                21                    -4                     1
 63   15   3   3            -670719                 30303               -239967
      -1   1  -1             41223                 -2127                 14229

819   42   5   1            -22890                 -6846                 -9072
273   d    o   2              1932                   358                   320
 63  -12   6   3    2173694690244210        -223628043696876      -910300082770260
      -1   1  -1     131570260248336         -13529445727800       -55089990435504

819   78   3   1             -4485                  1209                  -117
273   d    o   2               273                   -73                     7
117  -21   3   3             -4485                   585                   702
       0   0  -1               -39                   -27                    48

819   78   5   1            200694                -37440                -60762
273   d    o   2             11388                 -2096                 -3826
117    6  12   3  -73743890132694522      13669887390976908     23513394894063984
       1   2  -1  -4463183252437728        827339169423768      1423095406193952

819  546   5   1           1267539               -130494                  4641
273   d    o   2            -76713                  7898                  -281
819  -57   3   9          -14532609               -244881               1880424
      -1   1  -1          -1242969                 22605                112482

819  546   5   1              -546                  1911                  5187
273   d    o   2                 0                  -115                  -317
819   51  15   9           -670215                -73710                -10647
      -2  -1  -1            -39585                 -4596                  -873

819  546   5   1          -35177142               1575756              -2644278
273   d    o   2           -2133768                 95720               -159506
819   24  30   9      84501696275178          -3787910508564       6334839653688
      -2  -1  -1       5114572042944          -229276942248         383366455200

819  546   5   7              -546                     0                  2184
273   d   r1   2              1638                  -100                   -38
819   -3  33   9               546                  3276                  3276
      -2  -1  -1              2184                    72                   -60

824    1   6   3                48                    -5                    11
  8   d    r   1                36                    -3                     8
103  -13   9   1               -19                     0                    -4
      -1   0   0                -8                     2                    -2

824  103   5   1             84975                 -7931                 19158
824   n    o   1             -5926                   553                 -1336
103  -13   9   1           -419725               -121540               -34608
      -1   1  -1            -29258                 -8528                 -2614

828    1   5   1                37                   -15                   -44
 92   d    o   1                -8                     3                     9
  9   -3   3   1            256129                296838                196926
      -2  -1  -1             54924                 61294                 39344
```

829	1658	6	1	1067752	-63004	62175
829	n	o	1	-37296	2200	-2173
829	-7	33	1	-36739622	2085764	-2351044
	-1	0	-1	-1288540	70452	-82880
836	1	3	3	-63	50	68
44	d	o	1	12	-18	-22
19	-7	3	1	-63	50	68
	0	0	0	12	-18	-22
			b2	8	-4	-6
				-1	2	2
840	1	5	1	5	-22	7
120	d	o	2	6	0	3
7	-1	3	1	-7019	-7020	-2160
	0	3	0	1656	1074	654
840	7	5	1	35	-364	-581
840	d	o	4	-4	26	39
7	-1	3	1	-1693433	1565340	580860
	0	3	-2	-65834	-6632	-166510
844	211	5	1	102008583	-8822543	13842444
844	n	o	1	-7022561	607369	-952953
211	-13	15	1	-170671359	7350396	-58483714
	-1	1	-1	22806300	-2404202	1036962
849	566	5	1	-590781183094	103469067074	-13340036454
849	n	o	1	20275557720	-3551049190	457828866
283	32	6	1	-60515326522150	-10179938133720	-9015713164836
	-1	-2	-1	2076859396960	349378435128	309417779064
853	1706	4	1	-259725705	21027303	-9519480
853	n	o	1	10109439	-818455	370532
853	35	27	4	2370487	368496	199602
	-2	0	-1	-121053	-9384	-8310
855	2	6	1	58	6	-3
5	d	o	1	-30	-4	1
171	24	6	3	-58	0	-15
	-1	1	0	-30	2	-5
855	2	4	1	229	-24	30
5	d	o	1	-87	16	-10
171	-3	15	3	229	-24	30
	0	0	0	-87	16	-10
855	6	3	1	-1404	1065	-540
285	d	o	2	84	-61	34`
9	-3	3	1	-9939	10440	-17460
	0	0	-1	1635	-1032	-156
855	114	5	1	11799	4788	1596
285	d	o	2	-1197	-182	26
171	24	6	3	-5003346	-61560	541215
	-1	1	-1	-7410	67230	42453

855	114	3	1	-2052	-1653	-1140
285	d	o	2	-456	-55	-14
171	-3	15	3	-724413	-112518	-98838
	0	0	-1	-19779	-9930	-2994
860	43	3	1	5418	1935	1505
860	d	o	2	-325	-149	-95
43	8	6	1	178708	-69230	32035
	0	0	-1	12565	-4744	2019
861	14	3	1	5894	4725	2037
861	d	o	2	-200	-163	-73
7	-1	3	1	101003	81900	36582
	0	0	-1	-3475	-2756	-1222
868	1	3	1	8	-7	-14
28	d	o	1	-7	3	4
31	-4	6	1	8	-7	-14
	0	0	0	-7	3	4
868	7	5	1	1029	-21	-245
28	d	o	1	410	-4	-89
217	29	3	3	4669	1834	2744
	-1	-2	0	-5424	-638	-206
868	7	5	1	-2975	623	770
28	d	o	1	-1119	237	291
217	-25	9	3	-145397	-1134	-43470
	-2	-1	0	87978	-6372	7974
868	1	5	1	-61	-45	-23
124	d	o	1	11	8	4
7	-1	3	1	-108313	-86490	-38812
	1	2	-1	19350	15584	6878
868	31	5	1	-33170	496	7533
124	d	o	1	5957	-89	-1353
217	29	3	3	1039771	-22010	-249178
	-1	-2	-1	-198378	1828	42754
868	31	5	1	-9517	-1581	31
124	d	o	1	-1467	-363	-132
217	-25	9	3	22613413786753	-940318730928	3860985569166
	-2	-1	-1	-4061498931282	168884881770	-693453388476
869	158	5	1	124583	-5609	-50797
869	n	o	1	-4233	193	1723
79	17	3	1	-6444937473	-1235471204	-401045238
	-1	1	-1	-89909317	-47610812	-66217670
871	2	6	1	-239	-25	35
13	d	o	1	9	19	25
67	5	9	1	-49333778271	-17736645576	-10049057590
	-1	-1	-1	14384483823	4740486432	2911380438

871	26	6	1	585	−26	−91
13	d	o	1	209	−2	−21
871	53	15	3	−10303787	192634	1177800
	−1	−1	−1	−2859771	52634	326288
871	26	6	1	114517	12584	4472
13	d	o	1	28189	3106	1108
871	−28	30	3	8008	949	351
	−1	−1	−1	2434	249	85
872	1	4	1	−115	−26	−16
8	d	o	1	−61	−22	−10
109	2	12	1	7915	−594	1234
	0	0	−1	−3138	1118	−504
872	109	4	1	11445	−1962	1744
872	d	o	2	−761	130	−124
109	2	12	1	66163	−11990	−26814
	0	0	−1	4628	−892	−1792
873	2	6	1	28	−9	−25
97	d	o	1	−2	1	3
9	−3	3	1	875	−388	−1067
	−1	−1	0	−97	34	103
873	194	6	1	−76558414	−8642894	−2291334
97	d	o	1	7749524	879554	235362
873	42	24	3	1627456106	−135952484	−184806340
	0	−1	0	166073312	−13710288	−18739072
873	194	6	1	−21437	−2522	−1067
97	d	o	1	2425	282	121
873	15	33	3	−28033	1261	−2134
	−1	0	0	−3007	163	−176
876	1	3	1	11	−2	0
12	d	o	1	−2	0	−2
73	−7	9	1	11	−2	0
	0	0	0	−2	0	−2
876	73	3	1	54677	18250	7008
876	d	o	4	−3686	−1236	−478
73	−7	9	1	368869	122786	45698
	0	0	−2	−24422	−8430	−3224
877	1754	6	7	1927646	−149090	28941
877	n	r1	1	−65010	5026	−977
877	59	3	7	5262	7016	3508
	−1	2	−1	−2384	−12	−124
888	1	5	1	25	−6	−11
24	d	o	1	−4	−1	5
37	11	3	1	−1961843	−14568	868068
	−1	−2	−1	444266	−136582	−474452

```
888   37  5   1            28453           -14430              2701
888    d  o   2            -1908             969               -181
 37   11  3   1         25796881        -15173256           6382500
      -1 -2  -1         -2425506          1091182             33668

889   14  3   1       -175290122       -135938054         -61410349
889    n  o   1          5879042          4559216           2059637
  7   -1  3   1              406              168               175
       0  0  -1               -6              -10                -1

889  254  5   7   -17082533779492    4346522196194     -891527068352
889    n r1   1     572929794390     -145777675626       29900858184
127   20  6   1        22307042          5472684           4544568
       1  2  -1         -751704          -183072           -151800

889 1778  5   1  340554460746206   36775205692393    11161505412022
889    n  o   1  -11421830021970   -1233400812079     -374344876664
889  -37 27   3     -10136518462      -700095501       -457314046
      -1 -2  -1        153440904         29800929           833412

889 1778  5   1    1106379633947     120594359522       70389449137
889    n  o   1     -37106781955      -4044605004       -2360786353
889   17 33   3       -11939270         -1199261           -640080
       2  1  -1          357612            42397             26758

892  223  3   1           471645            -6913             79834
892    n  o   3          -31584              463              -5346
223  -28  6   1               0              223               446
       0  0  -3            -141               17                 6

893   38  5   1           319770           -47196            165129
893    n  o   1           -10710             1574             -5527
 19   -7  3   1   -1456742103019     214467586110     -752175837810
      -2 -1  -1      48748371987      -7177367190       25170022290

897    2  5   1             -708             -348              -154
 69    d  o   1              -66              -50               -48
 13    5  3   1          -150970           221352            535716
      -2 -1   0            55204            -4220            -48208

897   26  5   1            -3354             5421              -689
897    d  o   4              112             -181                23
 13    5  3   1            25142           -11661            -41262
      -2 -1  -2              930             -355             -1340

899    2  4   1              -19                3                -7
 29    d  o   1                3               -1                 1
 31   -4  6   1               55               -4                51
       0  0  -1              -19                6                -1

903    2  5   1              215              -82                36
 21    d  o   1               47              -18                 8
 43    8  6   1             -187                0               -42
      -2 -1   0               -1               16                 2
```

903	14	3	1	−546	0	−112
21	d	o	1	152	−8	16
301	−31	9	3	−546	0	−112
	0	0	0	152	−8	16
903	14	5	1	−12600	−1204	−1358
21	d	o	1	1378	458	218
301	23	15	3	15197714	−2166780	868056
	1	2	0	3324564	−471640	190108
903	2	3	1	−206	−90	33
129	d	o	1	18	8	−3
7	−1	3	1	314	−264	−39
	0	0	−1	18	−2	35
903	86	3	1	−22575	645	−3397
129	d	o	1	1987	−57	299
301	−31	9	3	5031	215	1677
	0	0	−1	−901	59	−53
903	86	5	1	−1290	−3311	−817
129	d	o	1	122	293	73
301	23	15	3	−7849564	484524	1632495
	1	2	−1	−673148	40034	144745
905	2	4	1	275	74	24
5	d	o	1	−153	−30	−16
181	−7	15	1	275	74	24
	0	0	0	−153	−30	−16
905	362	3	1	−520375	55929	−76201
905	d	o	4	17299	−1859	2533
181	−7	15	1	−136836	17195	−2896
	0	0	−2	1992	−131	858
909	2	6	1	318	−259	122
101	d	o	1	−34	23	−14
9	−3	3	1	817586114	−606871428	323025876
	−1	−1	−1	−81371256	60392300	−32122496
917	14	5	1	−1071	2359	3976
917	n	o	1	39	−75	−130
7	−1	3	1	−3905652457	−3129390880	−1394582770
	1	2	−1	−128844547	−103414440	−45961910
921	614	5	4	−67381965610	4675854372	−8245138296
921	n	o	1	2220311256	−154074640	271686544
307	−16	18	1	−587973154	82016892	128766852
	0	0	−1	−19425872	2707536	4236384
			c3	32541386	−2273642	3985474
				−1072012	74902	−131294
924	7	5	1	−539	−273	98
924	d	o	4	39	16	−4
7	−1	3	1	20294743	−9746814	18291966
	3	3	−2	1453194	−906202	726172

```
927    6   5   1          1005           1194            753
309    n   o   1           -57            -68            -43
  9   -3   3   1      -10147863      -11554128       -7532802
      -1   1  -1         578139         656616         429066

927  618   5   1     -3648423873     -370761375      -23717604
309    n   o   1      -402888105      -40942441       -2619092
927   60   6   3        -2290926         225261        -390267
      -1   1  -1          351642           9681          23643

927  618   5   1         7924305         891156         534261
309    n   o   1         -447741         -50834         -30739
927  -21  33   3     27979466415     7156541970     6482262564
      -1   1  -1     -4469407827     -277742802      -43600860

936    1   6   1              22             -1              8
  8    d   o   1              14             -7             -5
117  -21   3   3     -2118712439      558610104      -55208868
      -1  -2  -2     -1499977752      395044238      -38511158

936    1   4   1             -17             -6             -2
  8    d   o   1              14              4              2
117    6  12   3            -155             36            -30
       0   0  -2            -156             10            -28

936    3   5   1            -108              9             42
 24    d   o   1             -60             -1             13
117  -21   3   3        -5364213       -1550808       -1413252
      -1  -2  -1         2188056         633774         576690

936    3   5   4              -9              0             12
 24    d   o   1             -18              4              2
117    6  12   3            -933            324            432
       0   0  -1            -624             54            144
               c4             363            -72           -120
                              156            -27            -48

936    1   6   1              19            -24             19
104    d   o   2              -7              6              0
  9   -3   3   1        -1764983        2110212        3740568
      -2  -1  -2         -707928           2438         465058

936   13   6   1            2522            -65           -728
104    d   o   2            -494             13            143
117  -21   3   3         -263627         -77064         -70980
      -1  -2  -2          -52728         -15106         -13598

936   13   4   1              13            -78            -26
104    d   o   2              26             10             -4
117    6  12   3           -2015          -2340          -1794
       0   0  -2            1716            214            -70

936    3   5   1             -21              6             57
312    d   o   2               3              0             -6
  9   -3   3   1          424011         526500         322920
      -2  -1  -1           53352          55650          38670
```

```
936   39   5   1           2496              897              858
312    d   o   2           -312              -93              -99
117  -21   3   3    -75503449905      19895699304      -1953355716
      -1  -2  -1     -8549043048       2252799894       -221086110

936   39   5   4          10101            -1404             1872
312    d   o   2          -1170              150             -216
117    6  12   3        -182481            23868           -37908
       0   0  -1          22464            -3054             3702
              c3           -117             -156             -156
                             0              -21              -15

937 1874   6   4      109703960        -16202604        -20036808
937    n   o   1       -3583872           529316           654574
937  -61   3   4           1874             3748             7496
       2   0  -1           2452             -128              -12
             c-3          18740            -1874            -1874
                          -584               58               64

948    1   5   1            109              -39                4
 12    d   o   1            -64               22               -3
 79   17   3   1       10929943          -467490         -4492044
       1   2  -1        6373180          -291782         -2590810

949    2   6   1           1322             -685             -935
 13    d   o   1            588             -115             -231
 73   -7   9   1  -45176687587003  -15313004088434  -5778516818360
      -1  -1  -1   12529481591379    4247103832410    1602605126304

949   26   4   1          -8697             -858             -104
 13    d   o   1          -2661             -266              -34
949  -58  12  12           -143             -221             -130
      -2   0  -1           -625               -5               28

949   26   6   1            169               26               26
 13    d   o   1             35                8                6
949   23  33   3      -16147729          1089946          -567658
      -1  -1  -1        4001759          -280478           213526

949    2   4   1            -34               34               -9
 73    d   o   1              4               -4                1
 13    5   3   1            -11                7               31
       0   0  -1             -3                1                3

949  146   4   1        1287136           139576            23360
 73    d   o   1        -150554           -16340            -2724
949  -58  12  12      567324318         -54490996        -61918308
      -2   0  -1      -65629632          6367896          7314920

949  146   6   1         -30003             1898             4161
 73    d   o   1           3511             -222             -487
949   23  33   3      -24398498          1129018          2780351
      -1  -1  -1        2726260          -123204          -331069

949   26   4   1         -32773            13793            50414
949    d   o   2           1065             -447            -1636
 13    5   3   1        -222209          -139412          -106834
       0   0  -1          -7517            -4396            -3010
```

```
949  146   6   1        -6707240           983821         -1622863
949    d   o   2          217698           -31927            52695
 73   -7   9   1    4445427950681    -1692036569366    -3471964359968
      -1  -1  -1     246873578181      -69969405222      -87881980872

949  1898  4   1        -47490807        -4717479          -588380
949    d   o   2          1541649          153139            19100
949  -58  12  12            -1898             949             -949
      -2   0  -1             -652             -23              -35

949  1898  6   1          4927208         -224913          -578890
949    d   o   2          -159906            7305            18794
949   23  33   3        497051087        63398894         45326138
      -1  -1  -1         20904885         1839810           910382

952    1   5   1               24             -47              -87
136    d   o   2               -4               8               15
  7   -1   3   1             7549          -17408           -31348
       1   2  -2            -1354            2966             5348

952    7   5   1             2926            2233             1085
952    d   o   2             -190            -144              -69
  7   -1   3   1       -541611357      -434399504       -193400228
       1   2  -1         35109622        28151782         12523908

959    2   6   1              -18              11              -17
137    d   o   1                2              -1                1
  7   -1   3   1             -546             274             -411
       1   1   0               46             -28               29

963    6   3   1            34131          -25101            13491
321    n   o   3            -1905            1401             -753
  9   -3   3   1              105             -63                9
       0   0  -3               -3               3               -3

965    2   6   1              352             -69               18
  5    d   o   1              156             -31                8
193   23   9   1          6043717         -329190         -1579820
       0   1  -1         -2706257          146798           706036

965  386   6   1           -34354           -5597            -5211
965    d   o   2              898             221              157
193   23   9   1        -15374959          306870         -1547860
       0   1  -1           -62435           99438            21092

969   38   5   1           148732           86507            19019
969    d   o   2            -4778           -2779             -611
 19   -7   3   1          -999001          235467          -562989
      -1   1  -1           -39913            3037           -19093

973   14   3   1             -721           -5488            -7364
973    n   o   1               23             176              236
  7   -1   3   1              -98            2016             5488
       0   0  -1               72            -104             -128
```

```
973  278  5  1        -1539286          403934         -36140
973   n   o  1           49354          -12948           1160
139   23  3  1      -1538110338       403433044      -36347388
       1 -1 -1          49347684       -12936228        1150408


973 1946  5  1         25672605        -1475068       -2661155
973   n   o  1          -823069           47284          85311
973  -25 33  3       -5089850570       371351288      597386972
      -1  1 -1         164038556       -11813076      -19117136


973 1946  5  1      -14337675555       826742532     1684906153
973   n   o  1         458524169       -26431014      -53874973
973   2  36  3      174228356407     -9859048990     9602254830
      -2 -1 -1       -5623149711       318236910     -303410670


981   2   4  3              37             -30              6
109   d   o  1               3              -2              2
  9  -3   3  1              37             -30              6
      0   0  0               3              -2              2
            b1             -22              11             -9
                            -2               1             -1


981  218  6  3           69433           -1199          -7303
109   d   o  1           -6867              93            681
981  -57 15  3         -320569               0          41202
      0   1  0           44145           -1176          -3738
            b2           64092            -981          -6540
                         -6104              99            630


981  218  6  3          -37496           -4142          -1090
109   d   r  1           -3924            -376            -74
981   51 21  3          -35425            6758           6322
      1   1  0            7303            -234           -518


987   2   5  1              68             -55             53
141   d   o  1               8              -3              5
  7  -1   3  1           -5497            4794          -4230
      1   2  0            -679             230           -434


988   1   5  1             152             -65           -230
 76   d   o  1             -34              14             53
 13   5   3  1       -39464253         5610472       39455438
      -1 -2 -1         4140194        -4263054      -11212544


988  19   5  1            -133             -57            -38
 76   d   o  1              82               2             -3
247  -31  3  3        26129541        -5549862       -5904364
      -1  1 -1         6026680        -1267454       -1354070


988  19   3  1           65949           -6707           7752
 76   d   o  1          -15150            1533          -1782
247  -4  18  3         -810217           74993        -177973
      0   0 -1          289602          -31029          15081


988  13   5  1            1482             845            520
988   d   o  2             -94             -54            -33
 13   5   3  1          104247          -82992          36062
      -1 -2 -1            7006           -5426           1744
```

988	19	5	1		-20710	22135	28424
988	d	o	2		1315	-1408	-1810
19	-7	3	1		1524598133	-1642690790	-2105442820
	-2	-1	-1		-98177132	103845730	133815990
988	247	5	7		304057	57551	4446
988	d	r2	2		-19346	-3662	-283
247	-31	3	3		-247	-494	-988
	-1	1	-1		-316	34	6
988	247	3	1		332709	-20501	38532
988	d	o	2		-21354	1329	-2406
247	-4	18	3		-11893297	1006031	2359097
	0	0	-1		-534990	108123	170529
989	86	5	1		-12255	4386	-2064
989	n	o	1		391	-140	66
43	8	6	1		-667489	-284832	-186921
	-2	-1	-1		22701	8490	6207
993	662	5	3		122785112	21276018	10509912
993	n	o	3		-3896468	-675174	-333522
331	-1	21	1		4634	11916	0
	1	-1	-3		-2776	-132	-252
			b2		-3310	-3972	-3972
					84	128	124
995	2	6	12		277	-20	-150
5	d	o	1		-303	36	50
199	11	15	1		277	-20	-150
	0	0	0		-303	36	50
			b2		31	-5	-10
					-17	1	4
			c3		131	-14	-36
					-59	6	16
997							
997	n		1				
997	-10	36	1				
1001	2	5	1		46	33	17
77	d	o	1		-6	-3	-3
13	5	3	1		-9007	-5544	-4158
	-1	-2	0		1047	624	438
1001	14	5	1		-4494	-1253	-329
77	d	o	1		448	147	21
91	-16	6	3		-11844	-5929	539
	-2	-1	0		2814	581	315
1001	14	3	1		483	-112	84
77	d	o	1		-59	16	-4
91	11	9	3		483	-112	84
	0	0	0		-59	16	-4
1001	14	3	1		-6454	4081	-4620
1001	d	o	2		204	-129	146
7	-1	3	1		-294	693	1078
	0	0	-1		-8	17	36

```
1001   26  5  1              28431             -12441             -41262
1001    d  o  2               -899               393               1304
  13    5  3  1         -5491949437        -3326239917        -2415015603
       -1 -2 -1          -173581647         -105131649          -76343361

1001  182  5  1            -6002360           2072798            2595320
1001    d  o  2             189594            -65510            -82060
  91  -16  6  3 -1698730944981718466 586288895989568772 734490760572476908
       -2 -1 -1   53701626787038872 -18527838508812088 -23214430933881808

1001  182  3  1              26299             -4459             -5915
1001    d  o  2               -831               141               187
  91   11  9  3               3276               455               546
        0  0 -1                 44                29                12

1005  134  3  4               7504             -1005             -3015
1005    d  r  2               -226                29                97
  67    5  9  1               1407              -268              -402
        0  0 -1                -15                 4                18

1007    2  6  1              10251             -1462              5393
  53    d  o  1              -1423               216              -721
  19   -7  3  1      -2464295155263       2626008417602       3371173800944
       -1 -1 -1       -337086975963        360502405386        463794362488

1009 2018  4  1     -78168083122617       2285868369522     -5852618696658
1009    n  o  7      2460842841719        -71962399352        184248791190
1009  -43 27  4             -358195            -41369             -9081
       -2  0 -7              -13665             -1221              -423

1015    2  4  1                602               530               205
 145    d  o  4                -50               -44               -17
   7   -1  3  1               -218              -180               -95
        0  0 -4                 18                14                 5

1016    1  4  1                -21                 4                -2
   8    d  o  1                -12                 4                 0
 127   20  6  1                -21                 4                -2
        0  0  0                -12                 4                 0

1016  127  3  1            -8391271           2134616           -437134
1016    n  o  3             526514           -133938             27428
 127   20  6  1             -21209              3048             -1778
        0  0 -3                734              -334                -4

1017    2  6  1               -159               117               -64
 113    d  o  1                 15               -11                 6
   9   -3  3  1               5087             -1017            -10170
       -1  0 -1               1017              -495              -744

1021
1021    n     1
1021   14 36  1
```

1023	2	5	1	30558	-5906	10690
33	d	o	1	4732	-1310	1750
31	-4	6	1	-5722514950	-2758892796	-1077263220
	-2	-1	0	1002413872	478806680	189770872
1027	2	6	3	66	8	-27
13	d	o	1	22	-4	-9
79	17	3	1	-168751	7280	68770
	0	1	0	-46663	2120	19130
			b2	-647	-143	-78
				105	43	52
1027	26	6	3	-895050	-103129	-98475
13	d	o	1	333440	26951	18259
1027	56	18	3	-33332	1625	5109
	-1	-1	0	-14034	-9	1041
			b1	1014	78	39
				-176	-22	-25
1027	26	6	3	32864	3302	2093
13	d	o	1	-9060	-918	-587
1027	29	33	3	-817427	-73996	-41938
	0	-1	0	193507	21876	15362
			b2	-9529	-897	-598
				2429	263	160
1032	1	3	1	19	0	-6
24	d	o	1	-2	2	4
43	8	6	1	19	0	-6
	0	0	0	-2	2	4
1032	43	3	1	33583	13674	9288
1032	d	o	2	-2136	-834	-586
43	8	6	1	-1331237	512130	-223686
	0	0	-1	-83146	31604	-14270
1033	2066	6	1	-13845478742	1615638858	-158972502
1033	n	o	1	430782266	-50268292	4946202
1033	53	21	1	2373716238	230177192	181109692
	-1	0	-1	-74385768	-7118220	-5645908
1035	6	5	1	-1002	744	-408
345	d	o	2	54	-40	22
9	-3	3	1	49686	-37260	20700
	-1	1	-1	-2760	2028	-1032
1036	1	5	1	9	7	3
28	d	o	1	6	1	2
37	11	3	1	-3009	-1232	-1078
	-1	-2	0	-1184	-462	-374
1036	7	5	1	567	35	-28
28	d	o	1	-57	-39	-27
259	-19	15	3	-1505	-630	-378
	-2	-1	0	1290	120	-30
1036	7	3	1	10773	-1337	1141
28	d	o	1	-4374	543	-351
259	8	18	3	10773	-1337	1141
	0	0	0	-4374	543	-351

```
1036    7   5   1        -1946           1673            3976
1036    d   o   2          121           -104            -247
   7   -1   3   1         2079         515410         1448846
       -1   1  -1        38070         -53226          -63592

1036   37   5   1         9139          -2849           -2479
1036    d   o   2         -568            177             154
  37   11   3   1     -2338733        -939134         -787360
       -1  -2  -1      -145866         -58012          -49102

1036  259   5   1      -889147        -170681          -53872
1036    d   o   2        55167          10619            3367
 259  -19  15   3  -3351500663     -436563148       -27503210
       -2  -1  -1    135229598       39068078        19141136

1036  259   3   1       920227        -103341           85729
1036    d   o   2       -57138           6417           -5337
 259    8  18   3      7326592        -659414        -1644909
        0   0  -1       417813         -48144         -106341

1037    2   6   1           -5              0              -2
  17    d   o   1            1              0               0
  61   -1   9   1        -5455          -1921            -969
        0  -1  -1         1269            477             219

1037  122   6   1       -63562          13481          -14640
1037    d   o   2         1982           -421             450
  61   -1   9   1    261090797      -77463900      -149794650
        0  -1  -1      8147595       -2413500        -4642590

1043    2   6   1           -8            -68             -94
 149    d   o   1           -4              2               6
   7   -1   3   1        -5362           2384            7748
        0   1   0          -80           -612            -820

1044    3   5   1          -57            228             303
 348    d   o   2          -33             -9              12
   9   -3   3   1  -3808355397    -4335389568     -2832416370
       -1  -2  -1    408954636      464543490       302920638

1045   38   3   1        15143         -16530          -21185
1045    d   o   4         -473            512             653
  19   -7   3   1      -156427         165680          215080
        0   0  -2         4833          -5200           -6616

1055    2   4   1         -307             16             -90
   5    d   o   1         -233             24             -14
 211  -13  15   1         -307             16             -90
        0   0   0         -233             24             -14

1057   14   5   1    278300995     -155607886       192803457
1057    n   o   1     -8560069        4786236        -5930309
   7   -1   3   1     -2500848        5430866         9596503
        2   1  -1       -70056         163212          299929
```

```
1057  302   5   1     927468428020      -226794966571      -297355371672
1057    n   o   1     -28527363832         6975830477         9146149474
 151  -19   9   1     263936179116        64074222751        15201581626
        2   1  -1       8118634194         1970758185          467563404

1057 2114   5   1  -442377477237234     11275904383366     47120593319702
1057    n   o   1    13606784730320      -346827791826     -1449349939038
1057   50  24   3    -85724043136242     6763320369120     -2133503717492
       -2  -1  -1     2633208055200      -208367062528        65365291104

1057 2114   5   1     7067071982257       709027672500       251856716223
1057    n   o   1     -217371209169       -21808494790        -7746687607
1057  -31  33   3        229986288          26779095           7348264
        2   1  -1         -8734122           -762323           -337350

1064    1   6   1               69                 9                 11
   8    d   o   1              -25               -12                 -6
 133   17   9   3             7289             -1668                580
        1   2   0             5262             -1176                394

1064    1   4   1              -31               -10                 -4
   8    d   o   1               28                 6                  2
 133  -10  12   3              -31               -10                 -4
        0   0   0               28                 6                  2

1064    1   3   1               63                12                 -4
  56    d   o   1                6                10                  4
  19   -7   3   1             6813              2560               1612
        0   0  -1              804               834                -94

1064    7   5   1             1785               441                329
  56    d   o   1             -477              -118                -88
 133   17   9   3            20867              6328               4116
        1   2  -1            -6894             -1394              -1200

1064    7   5   4              483               336                252
  56    d   o   1             -370               -36                 18
 133  -10  12   3          -428141           -107604             -37296
        0   0  -1           109602             29838              11688
           c4                  119                14                 56
                               3               -3                 12

1064    1   5   1             -313              -205                -47
 152    d   o   1               46                44                 27
   7   -1   3   1     -16947138843      -13590484540        -6048224400
       -1   1  -1       2749185464        2204686210          981195550

1064   19   5   1             -247                57                -19
 152    d   o   1               13               -16                 -2
 133   17   9   3          -462289             50008             119700
        1   2  -1           -49206              2278              21408

1064   19   5   4              209               -76                -76
 152    d   o   1              -58                 6                 10
 133  -10  12   3            -7505             -1520               -836
        0   0  -1             -826              -288                -62
           c2                 -741               152               -152
                              185                -8                 31
```

```
1064    7   5   1            1001              -2233                  -4053
1064    d   o   2             -60                138                    249
   7   -1   3   1      -317343845         -254486988             -113226624
       -1   1  -1       -19458300          -15604530               -6946554

1064   19   3   1           -2793                304                  -1520
1064    d   o   2             162                -24                     92
  19   -7   3   1           48051              -8816                  25840
        0   0  -1           -3252                372                  -1616

1064  133   5   1          147763              36575                  27265
1064    d   o   2           -9063              -2242                  -1672
 133   17   9   3        -5973695            1354472                -454860
        1   2  -1         -366130              82574                 -28652

1064  133   3   1           29925               1330                   7980
1064    d   o   2           -2612                 92                   -214
 133  -10  12   3       -10747863             697718                2224026
        0   0  -1         -228470             155390                 177980

1069 2138   6   7    -69621830931         6105031206             -813902392
1069    n  r1   1      2129396819         -186723530               24893358
1069   62  12   1            2138              -1069                   1069
        1   1  -1            -692                 37                     25

1071    2   6   1              -4                 -2                     -2
  17    d   o   1               2                  0                      0
  63   15   3   3           11426              -4080                  -4692
       -1   1  -2            2856               -988                  -1112

1071    2   6   1          -57418               5728                  24466
  17    d   o   1          -14320               1516                   5894
  63  -12   6   3 -189523779547142038  19493452688868228   79362296869806216
        1   2  -2 -45966270691736976   4727855739146336   19248183528789448

1071    6   3   1            -342                189                    231
 357    d   o   2               6                 -1                    -17
   9   _3   3   1           26025              -9954                 -31248
        0   0  -1           -1491                606                   1608

1071   42   5   1           20979               7812                    672
 357    d   o   2           -1029               -442                    -68
  63   15   3   3     -11703319257        -4675791078             -569071566
       -1   1  -1       618543555          247510506               29814810

1071   42   5   1            -189                -21                     63
 357    d   o   2             -21                  1                      5
  63  -12   6   3        -7526946            2120580                -144585
        1   2  -1          270606             -99090                  61155

1073    2   6   4              63                -29                      0
  29    d   o   1              -9                  5                     -2
  37   11   3   1            1419               -116                   -986
        0   0  -1             285                -36                   -182
              c-4              46                 -5                    -34
                               10                 -1                     -6
```

1073	74	4	4	−22718	−8584	−7511
1073	d	r	2	694	262	229
37	11	3	1	−518	−185	−222
	0	0	−1	16	7	4
1084	271	5	1	−836765477248	142092382314	−30104173481
1084	n	o	1	50829882831	−8631497523	1828698306
271	29	9	1	−13952472605315	2369623137678	−502593227676
	1	2	−1	847857806850	−143955760896	30465147474
1085	2	6	1	−7976	150	1872
5	d	o	1	3700	−42	−814
217	29	3	3	13199502	−198820	−3001240
	1	−1	0	−5908032	88380	1341484
1085	2	4	1	−43	−4	2
5	d	o	1	9	4	2
217	−25	9	3	−43	−4	2
	0	0	0	9	4	2
1085	14	5	1	1288	−238	1841
1085	d	o	2	46	−40	3
7	−1	3	1	3914728839	−9979925480	−19166813610
	1	2	−1	−161320081	326549920	552489622
1085	62	5	1	806	−403	−248
1085	d	o	2	4	−3	−16
31	−4	6	1	424107032	203605675	80115315
	−2	−1	−1	−12881338	−6178803	−2430241
1085	434	5	1	59892	9765	7161
1085	d	o	2	1366	293	333
217	29	3	3	71353080029	−15870917790	672886620
	1	−1	−1	−2313011541	454639998	−45814836
1085	434	3	1	9114	217	1519
1085	d	o	2	−152	19	−41
217	−25	9	3	180327	−11718	33418
	0	0	−1	−6593	146	−1046
1092	1	3	1	52	−27	−30
12	d	o	1	−47	11	16
91	−16	6	3	52	−27	−30
	0	0	0	−47	11	16
1092	1	3	1	155	42	28
12	d	o	1	80	26	18
91	11	9	3	155	42	28
	0	0	0	80	26	18
1092	1	3	1	−1	18	30
156	d	o	2	2	−2	−4
7	−1	3	1	−1	18	30
	0	0	0	2	−2	−4

1092	13	3	1	338	−39	78
156	d	o	2	57	−1	16
91	−16	6	3	338	−39	78
	0	0	0	57	−1	16
1092	13	3	1	455	−78	52
156	d	o	2	−48	18	−6
91	11	9	3	455	−78	52
	0	0	0	−48	18	−6
1093						
1093	n		5			
1093	−22	36	1			
1099	2	6	1	49	−136	−243
157	d	o	1	−5	10	19
7	−1	3	1	−56219971	−45085690	−20058320
	1	1	−1	−4487171	−3598370	−1601960
1099	314	6	1	25748	−785	2355
157	d	o	1	−2138	55	−189
1099	−61	15	3	−48778801	4336654	5132330
	1	0	−1	−3994329	347566	401362
1099	314	6	1	−53225983	−5077223	−3722470
157	d	o	1	−4250553	−405147	−296762
1099	47	27	3	655633932594995605	−49265746102051234	17955224565661404
	0	1	−1	−52334704027708443	3930936377774358	−1433641149848364
1101	734	3	1	6953549	531416	−6606
1101	n	o	3	−209451	−16032	202
367	35	9	1	692162	−98356	17616
	0	0	−3	19712	−3052	532
1105	2	4	1	53	34	17
85	d	o	2	−5	−4	−3
13	5	3	1	869	578	476
	0	0	−2	−105	−58	−36
1105	26	4	1	−194064	−104312	−84201
1105	d	o	4	5838	3138	2533
13	5	3	1	−4173	−2431	−1547
	0	0	−2	119	75	61
1112	1	6	1	−955	253	−21
8	d	o	1	682	−177	17
139	23	3	1	−4851638675	1278428016	−122157364
	1	1	−1	3467684674	−904875854	75536700
1112	139	5	1	3871567	−1014978	90767
1112	n	o	1	−232146	60873	−5460
139	23	3	1	−32068595043401	770765064156	9381817814688
	2	1	−1	−1923333545976	46230802674	562686904830

1113	14	5	1	1470	35	3437
1113	d	o	2	-44	-1	-103
7	-1	3	1	476378	439635	155820
	2	1	-1	16824	11765	6430
1115	2	6	1	-2744	72	-483
5	d	o	1	-1238	32	-219
223	-28	6	1	-10333	2195	2480
	-1	0	-1	4501	-973	-1134
1116	3	5	1	2760	513	30
12	d	o	1	-1605	-295	-17
279	33	3	3	35129271675	6491236284	378412902
	2	1	-1	-20286786756	-3747663438	-219326166
1116	3	5	1	528	-33	-108
12	d	o	1	-303	19	62
279	-21	15	3	-93741	5076	20934
	-2	-1	-1	62868	-4170	-11556
1116	1	5	1	-4	-5	-7
124	d	o	1	-1	-1	0
9	-3	3	1	-743	-806	-496
	-2	-1	0	-124	-148	-102
1116	31	5	7	11501	2077	155
124	d	r2	1	-1984	-375	-16
279	33	3	3	31	-62	62
	2	1	0	124	10	12
1116	31	5	1	57691	-3503	-11656
124	d	o	1	10354	-629	-2094
279	-21	15	3	-46097	22010	-53258
	-2	-1	0	77252	-8140	-4374
1117	2234	6	1	144645582134	-12683074796	1206151121
1117	n	o	1	-4327966044	379492524	-36089461
1117	65	9	1	44321735284273	3912899053340	3573193517130
	1	1	-1	-1326207964263	-117076511820	-106906599990
1121	38	5	1	-30282732	-17889849	-4236829
1121	n	o	1	904466	534323	126543
19	-7	3	1	-134599553	20647699	-67794717
	-1	1	-1	-3966275	559135	-2098799
1129	2258	6	1	-33817923409558	87438807806	-2891344609084
1129	n	o	9	1006467985916	-2602299370	86050398484
1129	-67	3	7	2258	-4516	4516
	-3	-1	-9	-3100	-128	-140
1132	283	3	7	-3949831	101314	784193
1132	n	r1	1	234796	-6022	-46615
283	32	6	1	566	566	283
	0	0	-1	197	6	19

```
1133   206   5   1        -1445811            135651           -324347
1133    n    o   1           43019             -4049              9609
 103   -13   9   1      -7182917283        2067940270        2929178890
       -2  -1  -1       -212255971          61328470          87277690

1137   758   5   1      -35416944358       4669574483       -1563926445
1137    n    o   1       1050342302        -138483195          46380571
 379    29  15   1         -1093036          -156906            753831
       -1  -2  -1          -175990            14280             16017

1139    2    6   1               29               -6               -18
  17    d    o   1               -9                2                 4
  67    5    9   1              138              -34               -85
       -1  -1   0              -42                8                19

1140    1    5   1              -81                8               -17
  60    d    o   2               -7                2               -10
  19   -7    3   1       3443499751       -3677053650       -4726288260
        0    3  -2       -889658456         949491694        1220038358

1141   14    5   1            53032           -82488           -111909
1141    n    o   1            -1570             2442              3313
   7   -1    3   1         -95934125         -75000212         -34716066
        1    2  -1          -2755729          -2267148           -969370

1141   326   3   1        838280149         -14609690         176420768
1141    n    o   1        -24816833           432512          -5222842
 163   -25   3   4         39192535          -654934           8315608
       -2    2  -1         -1167899            20950           -244272

1141  2282   5   1     -12614866885103     518473724874     1345660298276
1141    n    o   1      373450845771      -15348909432      -39836962132
1141    26  36   3       6537028610        -412582177         263711343
       -2  -1  -1       -195816036          12307893          -7563309

1141  2282   5   1       -446520081         -45514490         -23602726
1141    n    o   1         13218987          1347432            698746
1141   -1   39   3       96055549387       -4881551710       4979682274
        1    2  -1       2848778151        -144790214         146872818

1143    6    3   1             6729            -2664             -7656
 381    n    o   1             -345              136               392
   9   -3    3   1           -17274           -29376            -24768
        0    0  -1            -1536            -1248              -528

1143   762   5   1          2758821           -71247           -294132
 381    n    o   1          -141351             3649             15068
1143   -57  21   3       10012625898       -191425068        -972450684
        2    1  -1        -515883144          10032588          49763568

1143   762   5  16        132387594          -6611874         -13567410
 381    n    r   1         -6784848           338488            694952
1143   -3   39  12             -762            -4572             -4572
       -2    2  -1            -3048              -84                72
            c-4             14478             -762             -1524
                            -762               36                78
```

```
1144    1  5  1              -3              -1              -3
  88    d  o  1               1               0               0
  13    5  3  1             -87              44             132
       -2 -1  0             -18               6              28

1144   13  5  1           34853           21723           16549
1144    d  o  2           -2059           -1286            -978
  13    5  3  1      -1519084983      1232372856      -650414908
       -2 -1 -1    .  -102775588        78319406       -18569890

1145    2  6  1            2748            -512            -686
   5    d  o  1           -1210             228             310
 229  -22 12  1           -1258             240             340
        0  1  0             608            -112            -144

1145  458  6  1     -3008719248       563845632       761245006
1145    d  o  4        88960358       -16665624       -22489818
 229  -22 12  1 -28200486123704593202 5221670043988302020 6938630963182055120
        0  1 -2  805544325678166936 -152776897441469592 -209451153356126720

1147    2  6  1              -4               3               9
  37    d  o  1              -2               1               1
  31   -4  6  1           -6325           -1850              74
       -1 -1 -1             665             514             334

1147   74  4  1       780755685        73911681        19270488
  37    d  o  1      -128353093       -12151065        -3167880
1147  -49 27 12      5102239838      -128910220       366050324
       -2  0 -1       838252380       -21246852        60163200

1147   74  6  1            3848            -296            -518
  37    d  o  1            -622              48              84
1147    5 39  3        -1334146           70448            1036
       -1 -1 -1          -79948            4396          -14840

1148    7  3  1            1876            1526             938
1148    d  o  2            -117             -76             -30
   7   -1  3  1         -261807         -196294          -73668
        0  0 -1           14926           12776            6494

1153 2306  6  1     28852379138     -1384943245      1557685705
1153    n  o  1      -849702658        40786583       -45873849
1153   -7 39  1       -39635528        -3975544        -1938193
        0  1 -1        -1215646         -114314          -51717

1155    2  5  1              88            -351            -754
 165    d  o  2              20             -33             -50
   7   -1  3  1          533777        -1203840        -2171070
        3  3  0           41885          -93768         -168822

1157    2  6  1             771            -324           -1177
  89    d  o  1             -81              34             125
  13    5  3  1       -301464625       171942037       549881961
       -1 -1 -1        41276769       -12580115       -54187973
```

```
1157    26   6   1         -4238              -2730              -2054
1157    d    o   2           130                 78                 52
  13    5    3   1      9552178142         5786212536         4202395236
       -1   -1  -1      -280873164         -170088516         -123474312

1159     2   6   1            10                -12                -20
  61    d    o   1            -2                  2                  2
  19   -7    3   1          -730                732                976
       -1   -1   0            88               -100               -124

1159   122   6   1           183               4514               6161
  61    d    o   1         12365                584                 33
1159    44  30   3         14213               -244              -2135
         0   1   0         -3095                118                235

1159   122   6   1         77348               7381               2440
  61    d    o   1         -9876               -947               -316
1159   -37  33   3        434503              43066              13542
       -1   -1   0        -58575              -5410              -1894

1164     1   5   3          -180                -53                 -5
  12    d    r   1           108                 30                  3
  97   -19   3   1           163                 54                  0
       -1   -2   0          -120                -30                 -6

1164    97   5   1        834006             -23959             221063
1164    d    o   4        -48892               1404             -12959
  97   -19   3   1     -337032029           28513926          -51665304
       -1   -2  -2      13113740             725002            5685494

1169    14   5   4      23636025           19137699            8394589
1169    n    o   1       -691301            -559735            -245523
   7   -1    3   1        -22197             -17535              -8183
        0    0  -1           635                517                219
                 c4          -259               -140                -84
                               5                  6                  2

1179     6   5   1          9099              -3132              -9714
 393    n    o   1          -459                158                490
   9   -3    3   1      -1555095             609543            1763784
       -1    1  -1         78207             -31119             -89166

1185   158   5   1        135801             -20619             -34286
1185    d    o   2         -3945                599                996
  79    17   3   1      -5968687           -1709955           -1499025
       -1    1  -1       -169517             -49819             -45167

1196     1   5   1           -37                -10                -15
  92    d    o   1             4                  5                  2
  13    5    3   1           323                598                184
       -2   -1   0          -188                -30                -74

1196    13   5   1        -28223              10894              41457
1196    d    o   2          1522               -697              -2446
  13    5    3   1  -227810974324063  -137977913733374  -100189571378078
       -2   -1  -1   -13174884271514    -7979369733732    -5793750860630
```

```
1197    6   5   1          -957              -150              18
  21    d   o   1          -129               -50             -16
 171   24   6   3        -71814            -19341           -1512
      -1   1  -1        -18354             -4125            -894

1197    6   5   1        -10191             -2763           -1446
  21    d   o   1         -2295              -593            -322
 171   -3  15   3  -50439860170743    7119324329670  -6232696711164
      -1   1  -1   11006863027125   -1553568608154   1360083325308

1197   42   5   1         -8862               294            -189
  21    d   o   1           504                58             167
1197   69   3   9     -818133897          69841044        71920422
      -1  -2  -1     -178405269          15252396        15693966

1197   42   3   1       -1239609            100632          -12600
  21    d   o   1        -236313             19184           -2402
1197  -66  12  36             42              -315            -126
       2   4  -1            798                 3              36

1197   42   5   1          50757             -3360            1617
  21    d   o   1          11067              -734             353
1197  -39  33   9      -11940033           1194354        -1905246
      -2  -1  -1       -7723443            423774           86502

1197   42   5   1       -9551577            416619         -537180
  21    d   o   1        2084271            -90919          117220
1197   15  39   9  -43443806610729    1895007966516  -2443235342562
       1  -1  -1    9480080117109    -413516941932    533171274354

1197    6   5   1            -60               -21              -6
  57    d   o   1              6                 3               0
  63   15   3   3           -507                 0            -171
      -2  -1   0            -57                  6             -21

1197    6   5   1          31818             -7914            1920
  57    d   o   1          -2808              1170            -468
  63  -12   6   3  -172839832770       54591020172    -17759013480
      -1   1   0    22883846544       -7229795040      2356154424

1197  114   5   1         -13110                57           -1083
  57    d   o   1           1710                -3             147
1197   69   3   9        -419007            -12825          -23085
      -1  -2   0          14877             -1875            2955

1197  114   3   1       19727814          -1584372          176016
  57    d   o   1        -2539920           208464          -29064
1197  -66  12  36        2261190            -13680         -190836
       2   4   0         275424             -4032          -27240

1197  114   5   1           9633              -627             285
  57    d   o   1           1311               -87              45
1197  -39  33   9        -146091              5301           15048
      -2  -1   0          20463              -567           -1914
```

136

```
1197  114  5   1         13110           -570              741
  57    d  o   1           684            -30               39
1197   15 39   9         -9633           -513                0
        1 -1   0           513            111               78

1197    2  3   1          -107            -98              -70
 133    d  o   1            -7            -10               -6
   9   -3  3   1          -107            -98              -70
        0  0   0            -7            -10               -6

1197   14  5   1          1456            882              252
 133    d  o   1          -196            -52                6
  63   15  3   3        625646         -238336          -261212
       -2 -1   0         58520          -18964           -22440

1197   14  5   1         -3220          -1351            -1022
 133    d  o   1           280            117               88
  63  -12  6   3          5600           2527             1862
       -1  1   0          -532           -207             -160

1197   38  5   1          8702           2242              380
 133    d  o   1           760            194               32
 171   24  6   3        -60610            532           -11704
       -1  1   0          4256           -296              976

1197   38  5   1         -3287           -190              247
 133    d  o   1           -19             54               59
 171   -3 15   3         -7543           6916             8246
       -1  1   0         -3325           -108              338

1197  266  5   7       -424270          36442            37772
 133    d r1   1         37240          -3164            -3234
1197   69  3   9           266           1064              532
       -1 -2   0          1064              4              -44

1197  266  3   1      73444462        -5962257          746529
 133    d  o   1      -6359794         516293           -64647
1197  -66 12  36          -266            133             -133
        2  4   0           266            -13               -9

1197  266  5   1           266           1197            -2394
 133    d  o   1         -3458            219              118
1197  -39 33   9     -737896369       26563558         76261136
       -2 -1   0      68040539        -1905186         -6343864

1197  266  5   1        -53333           2261            -3059
 133    d  o   1          3857           -173              223
1197   15 39   9      -4722697         -416290         -155078
        1 -1   0       -348061          -39554          -19598

1201 2402  6   1   -373513707054    -1979193955     -16962108521
1201    n  o   1     10777922100       57110617        489450001
1201   59 21   1     -276369316        5122265         26875978
       -1 -1  -1       7912504         -154327          -780304
```

1204	1	3	1	27	-7	-14
28	d	o	1	-6	1	6
43	8	6	1	771	-161	-539
	0	0	-1	-314	63	199
1204	7	5	1	3171	-245	616
28	d	o	1	-1298	107	-214
301	-31	9	3	965107847193189	-169247095463670	-202886154431240
	1	-1	-1	364777559032314	-63969201767460	-76683727240170
1204	7	5	1	1295	224	161
28	d	o	1	-492	-85	-61
301	23	15	3	113001	27398	25032
	1	2	-1	-67378	-8860	-4246
1204	1	3	1	-15	40	88
172	d	o	1	-4	8	12
7	-1	3	1	-15	40	88
	0	0	0	-4	8	12
1204	43	5	1	-5676	-2021	-344
172	d	o	1	1397	211	-62
301	-31	9	3	-599495551	104724608	125023274
	1	-1	0	-90571116	15944994	19192926
1204	43	5	1	19780	3397	2451
172	d	o	1	2965	526	371
301	23	15	3	-7611	-6192	-1978
	1	2	0	-5844	-278	-570
1205	2	4	1	-327	16	-30
5	d	o	1	-61	24	-2
241	17	15	1	-327	16	-30
	0	0	0	-61	24	-2
1205	482	3	1	976532	-82181	-242205
1205	d	o	2	-28946	2209	6871
241	17	15	1	-193070643	-37602266	-25096776
	0	0	-1	-5543359	-1084578	-728096
1208	1	6	1	13	-4	-5
8	d	o	1	10	-2	-3
151	-19	9	1	-475	-108	-32
	1	1	0	300	78	14
1208	151	5	1	-806189	196753	258059
1208	n	o	1	46391	-11322	-14850
151	-19	9	1	-9471173	591920	-1827100
	2	1	-1	-537758	32110	-107600
1209	2	3	1	13	3	-24
93	d	o	1	-3	1	2
13	5	3	1	-517	252	648
	0	0	-1	37	-12	-72

```
1209   62  3  1              -3317                    31                    -558
  93   d   o  1                423                   -15                      44
 403  -37  9  3            -131223                 20212                   23188
       0   0 -1             -13689                  2044                    2428

1209   62  5  1              16492                  1333                    1240
  93   d   o  1                896                   227                      74
 403   17 21  3    -628058999104147        68221882719270         -40509857201526
        1  2 -1      65075887946427         -7070771765286           4210083990414

1209   26  3  1             105014                 42276                    5148
1209   d   o  2              -3020                 -1216                    -148
  13    5  3  1              59774                -40248                  -15756
        0  0 -1                668                  -716                    1160

1209   62  5  1            -4305342              -1973398                 -836938
1209   d   o  2             123908                 56706                   23990
  31   -4  6  1  8357658954011887946   4010638675484471160   1577788934673705156
       -2 -1 -1 -240365150150911720  -115345426725779152   -4537699440543016

1209  806  3  1            -4775147                109213                 -632307
1209   d   o  2             137333                 -3141                   18185
 403  -37  9  3            -246636                 34255                   35464
        0  0 -1              -5592                   939                    1226

1209  806  5  1             399373                 43927                   19747
1209   d   o  2             -11491                 -1263                    -567
 403   17 21  3          -10923718                766506                 2027493
        1  2 -1            -314818                 21392                   58177

1211    2  6  7             -11395                 -8828                   -3616
 173   d r2  1                851                   706                     338
   7   -1  3  1            -147913                 84078                 -102762
        0  1 -1             -11463                  6210                   -7902

1213 2426  6  1            -339640                 31538                  -48520
1213    n  o  1               9866                  -912                    1398
1213   17 39  1         -3471079558             163138796               370600612
        0  1 -1          103109444              -4359112               -10455268

1221    2  5  4                 64                   -33                       0
  33   d   o  1                -10                     5                      -2
  37   11  3  1                 64                   -33                       0
        0  0  0                -10                     5                      -2
           c-3                   3                     2                       0
                                1                     0                       0

1221   74  5  4              -1295                 -2442                   -1221
1221   d   o  4                 55                    68                      23
  37   11  3  1             -57313                -51282                  -75702
        0  0 -2               4861                  1130                      22
            c2               2257                    37                    -962
                               33                   -11                     -38

1224    1  3  1                -23                    40                      -4
 136   d   o  2                 -8                     2                      -4
   9   -3  3  1                -23                    40                      -4
        0  0  0                 -8                     2                      -4
```

1224	3	3	1	1257	-492	-1440
408	d	o	2	-126	50	142
9	-3	3	1	-19653	9252	24156
	0	0	-1	2136	-696	-2250
1228	307	3	1	-286834091	19919388	-35101459
1228	n	o	1	16370418	-1136850	2003361
307	-16	18	1	3292708238	-459261869	-719907018
	0	0	-1	188037621	-26210589	-41075466
1233	2	6	1	-1450	573	1651
137	d	o	1	124	-49	-141
9	-3	3	1	-7695151	715140	5899905
	-1	-1	-1	398259	-355830	-696435
1235	2	6	1	-38	-47	31
5	d	o	1	194	19	17
247	-31	3	3	-1574533	17890	-274890
	-2	-1	0	-704193	8242	-122786
1235	2	6	1	-1789	252	458
5	d	o	1	-817	112	206
247	-4	18	3	-273	30	-10
	-2	-1	0	69	-6	18
1235	2	6	1	-251	49	-119
65	d	o	2	33	-5	15
19	-7	3	1	-21206768	867880	-9744345
	-2	-1	-2	1859686	-553418	1109731
1235	26	6	1	-18226	-3354	-208
65	d	o	2	2262	416	26
247	-31	3	3	4708366	-57460	824720
	-2	-1	-2	588536	-6396	102284
1235	26	6	1	-1069302	205686	330174
65	d	o	2	179832	-16126	-36606
247	-4	18	3	-2139623228314	-351500493240	-203337854460
	-2	-1	-2	-204273852696	-49797796872	-18034681776
1236	1	3	1	-142	-88	-52
12	d	o	1	175	24	-8
103	-13	9	1	-132591	-38568	-11480
	0	0	-1	76604	22228	6528
1237	2474	6	1	-4940578	389655	-141018
1237	n	o	1	140720	-11095	4016
1237	41	33	1	6951157722437	635846337196	433693869950
	-1	0	-1	-197722513625	-18073141300	-12333583702
1239	2	5	1	-6	1	13
177	d	o	1	0	-1	-1
7	-1	3	1	-352	531	1062
	2	1	0	14	-51	-84

```
1240    1   4   1            9                 0                10
  40    d   o   2           -6                 2                 0
  31   -4   6   1            9                 0                10
        0   0   0           -6                 2                 0

1240   31   5   4        13671             -8060            -14260
1240    d   o   2         -776               458               810
  31   -4   6   1        16771              8680              3100
        0   0  -1         1050               468               202
                c2        1457               682               248
                            81                40                17

1241    2   4   1           13                -1                -3
  17    d   o   1           -1                 1                 1
  73   -7   9   1          141                43                21
        0   0  -1           29                11                 3

1241  146   4   1      -108259             -5037             17155
1241    d   o   2         3073               143              -487
  73   -7   9   1         7811             -1825              -949
        0   0  -1            9                17                83

1247    2   6   1           -7                 3                -4
  29    d   o   1           -3                 1                 0
  43    8   6   1         1452              2871              3973
       -1   0  -1         1678               245              -199

1249 2498   6   1     90399997100      -5749028345        1647588374
1249    n   o   1     -2557921412        162672159         -46619488
1249   53  27   1   -364158757246       8238932327       34764276312
        1   0  -1     9991386414        -261027153       -1004642982

1251    6   5   1        -25791             23892             -9312
 417    n   o   1          1263             -1170               456
   9   -3   3   1      -6218715           1977831           6485184
       -1   1  -1       -273969            131613            340266

1251  834   5   1    16869088650        -473687814       -1635020304
 417    n   o   1       588980808        -14437974         -54264156
1251  -48  30   3   -885652845078       8644975452       64044389556
       -2  -1  -1    -31354603632       1564020792        3957368112

1251  834   5   1      -4230465            170970            403239
 417    n   o   1        207249             -8376            -19755
1251  -21  39   3    -98022793467       5641922430       -4201089435
        1  -1  -1      4984475643        -283697160         187770795

1253   14   5   1        160741            -89614            110054
1253    n   o   1         -4527              2500             -3166
   7   -1   3   1 -245017286282755   550548387832142   992053415112410
        2   1  -1   -6921818427527    15553197737814    28025908050122

1256    1   4   1         -1073              -298              -176
   8    d   o   1          -909              -182              -140
 157   14  12   1       -647017            132870            -57410
        0   0  -1        494012            -85360             46380
```

1256	157	4	1	28103	-5338	2512
1256	d	o	2	-1571	300	-146
157	14	12	1	144911	-15386	-42390
	0	0	-1	7530	-712	-2446
1260	3	3	1	219	-54	108
60	d	o	2	-108	-6	-30
63	15	3	3	219	-54	108
	0	0	0	-108	-6	-30
1260	3	3	1	-93	-90	-45
60	d	o	2	-60	-12	-15
63	-12	6	3	-93	-90	-45
	0	0	0	-60	-12	-15
1260	1	5	1	35	-44	77
140	d	o	2	-20	13	3
9	-3	3	1	-101725679	32791290	106621200
	-3	-3	-2	-15565620	7395352	19231478
1260	7	3	1	-679	-266	-28
140	d	o	2	112	46	6
63	15	3	3	4711	2058	126
	0	0	-2	-924	-340	-62
1260	7	3	1	-497	70	245
140	d	o	2	98	-6	-37
63	-12	6	3	-19103	-8295	-6090
	0	0	-2	-3360	-1357	-1052
1261	2	6	1	21	-41	36
13	d	o	1	77	9	12
97	-19	3	1	797318953	-22967854	211383952
	0	-1	-1	221159625	-6365726	58626552
1261	26	6	7	-325	-65	-26
13	d	r2	1	-77	-15	-8
1261	-61	21	3	-5317	78	286
	0	1	-1	-347	122	98
1261	26	6	1	319111	29614	10595
13	d	o	1	737673	68066	24159
1261	-34	36	21	467925484637	-30674758361	-47525068448
	-2	-3	-1	129783246573	-8507762223	-13180837302
1261	2	6	1	218	127	89
97	d	o	1	-22	-13	-9
13	5	3	1	-189342	147828	-56357
	0	-1	-1	-19124	15094	-5633
1261	194	6	1	-1455	873	679
97	d	o	1	979	-15	-55
1261	-61	21	3	13531403	-1099689	1451023
	0	1	-1	-3043555	-36269	-175343

```
1261  194   6   1      -1248609802       -115138612        -40815078
  97    d   o   1       -126816552        -11691212         -4146114
1261  -34  36  21
       -2  -3  -1

1261   26   6   1          -243061           190801           -71981
1261    d   o   2             6845            -5373             2027
  13    5   3   1      -1606671599       1612476008      -2241899114
        0  -1  -1         99542569        -68249344        -20261298

1261  194   6   1            66057           -23765           -26190
1261    d   o   2            -1859              669              738
  97  -19   3   1       -373519355        -20909902         46339228
        0  -1  -1         -3043323         -3285406         -1684524

1261 2522   6   1            11349            -8827            -2522
1261    d   o   2            -3363              -21               20
1261  -61  21   3      61092992311      -1045391698       4464184634
        0   1  -1      -1721985865         29308582       -125734710

1261 2522   6   1    -137104018142     -12640844060      -4481759191
1261    d   o   2       3860943012        355974832        126209415
1261  -34  36  21      -3510148603       -341444753       -129903176
       -2  -3  -1        105207429          9198393          3012162

1267    2   6   4            -1305              724             -905
 181    d   o   1               97              -54               67
   7   -1   3   1             -545             2896             6878
        0   0  -1             -187              296              410
              c-4              -17              -11              -66
                                3               -3               -2

1267  362   6   1            -4344             -362             -543
 181    d   o   1              558               26               19
1267   71   3   9       -216029895           502094         18978212
       -2  -1  -1        -16028815            41006          1413516

1267  362   4   1      -1534236183       -134356662        -21713484
 181    d   o   1       -114985857        -10069580         -1627352
1267  -64  18  36          2494180           -37829           182991
       -2  -4  -1          -183714             2487           -13887

1271    2   6   7          -140912            83532           148250
  41    d   r¹  1           -26454            14076            21558
  31   -4   6   1            -8198             4100             7052
       -1   0   0            -1136              712             1128

1273   38   5   1        106140118         -7455296         50264538
1273    n   o   1         -2974852           208954         -1408794
  19   -7   3   1           -45790           -81472           -56012
       -2  -1  -1             3156              284            -1000

1273  134   5   1        976281455       -211418835       -512705038
1273    n   o   1        -27362819          5925561         14369888
  67    5   9   1      -1817227734        394394495        894554922
       -2  -1  -1         46214196         -9851901        -25907784
```

```
1273 2546  5  1      1476148203557488       131296924882956         28431676226974
1273   n   o  1       -41372880663682        -3679936737788          -796871442022
1273 -58  24  3   128496282822050542694  -9828413597119872428  -12544980627898008880
      -1  -2 -1   3601444732366881376   -2754664844481613448    -351605545295090248

1273 2546  5  1         -413774010500          -37977014370           -22596378862
1273   n   o  1          11597089450           1064404292              633322104
1273  23  39  3         -20781035034          -1909347240            -1135368332
       1  -1 -1            583060360             53478052               31851292

1281   2   5  4                   79                   -21                      0
  21   d   o  1                    9                    -1                      6
  61  -1   9  1                 3509                 -1092                  -2478
   0   0  -1                 -1023                   292                    482
          c2                   192                   -58                   -121
                             -48                    14                     25

1281  14   5  1               -25144                   175                   3913
  21   d   o  1                -5444                    43                    859
 427  41   3  3           -21128749243           -2886457938           -2774861838
      -1   1 -1            4684237415             629330354              593996518

1281  14   5  1                -8785                  -770                   -490
  21   d   o  1                 2617                   484                    -46
 427 -40   6  3            -793483936             120790425              133262325
       1   2 -1             173135570             -26363245             -29078795

1281  14   5  1                43057                 42448                  26593
1281   d   o  2                -1203                 -1186                   -743
   7  -1   3  1             -2659342               1456497              -1893318
       1   2 -1               -74912                 42113                 -50360

1281 122   5  4              1258918               -398391               -748104
1281   d   o  2               -35174                 11131                  20902
  61  -1   9  1               100040                -30744                -60207
   0   0  -1                 -3010                   862                   1637
          c2                  8357                 -2440                  -4575
                             -219                    66                    131

1281 854   5  1             -8344861                134932               1453081
1281   d   o  2               233155                 -3770                -40599
 427  41   3  3            -29441650               4546269               -432978
      -1   1 -1              -895778                127595                  -572

1281 854   5  1              5632130               -118706               -491050
1281   d   o  2              -160752                  2734                 13746
 427 -40   6  3  -329708811013463891392 5003200981101202644240 55273016859564720745240
       1   2 -1   9180764714249759380800 -14023891219800450040 -15477451492282183040

1287   6   5  1                  -54                    12                      3
  33   d   o  1                   12                    -2                     -1
 117 -21   3  3              -415695                110781                -12474
      -1  -2 -1                74217                -19341                  1620

1287   6   3  1               244242                -38388                -11124
  33   d   o  1                -3576                  -536                 -10480
 117   6  12  3           14808039486            -2742771168            -4719607092
   0   0  -1              2576234544             -477937464             -821780328
```

1287	6	5	1	−1089	−1299	−894
429	d	o	2	57	61	38
9	−3	3	1	12283247697	13970408166	9117422028
	−2	−1	−1	−593165001	−674405790	−440240916
1287	78	5	1	585	−1131	1326
429	d	o	2	−429	65	52
117	−21	3	3	14198813421	−367078140	−4109233986
	−1	−2	−1	685641957	−17743908	−198388554
1287	78	3	1	2457	−312	468
429	d	o	2	−117	16	−22
117	6	12	3	3120	−351	1404
	0	0	−1	−312	45	−18
1288	1	5	1	9	−7	−17
184	d	o	1	0	2	3
7	−1	3	1	−91	−368	−460
	−1	1	0	−42	10	48
1288	7	5	1	13783	−7679	9457
1288	d	o	4	−768	428	−527
7	−1	3	1	120366827	−66570924	83936384
	−1	1	−2	−6722880	3743602	−4616714
1292	19	5	1	39805	−513	18069
1292	d	o	4	−1417	490	−903
19	−7	3	1	−55846062741007	8647913465700	−29062845133330
	−1	1	−2	3172655470434	−443414476750	1625479846080
1295	2	4	1	828	123	63
5	d	o	1	242	63	9
259	−19	15	3	268175	−19350	38322
	0	0	−2	−126963	7302	−17562
1295	2	6	1	9451	−960	−2252
5	d	o	1	−4221	434	1012
259	8	18	3	−2848	675	1135
	1	2	−2	2438	−77	−379
1295	2	6	1	−184	−330	−322
185	d	o	2	14	24	24
7	−1	3	1	23454302	−11398960	20762180
	−1	1	−2	1864936	−1153860	957428
1295	74	4	1	−15466	−2886	−851
185	d	o	2	1140	212	63
259	−19	15	3	−22755	1665	−3071
	0	0	−2	−1691	95	−247
1295	74	6	1	−1464312	311540	541754
185	d	o	2	−188486	7324	30910
259	8	18	3	180025094774	244579626660	268465081000
	1	2	−2	−117884097560	−7299885352	5259610912

```
1297 2594  6  1    -1056403554513      40025844119      -60465846878
1297   n   o 11       29333228531      -1111400305        1678959236
1297 -25  39  1          11827343         -546037             693895
       -1   0-11            371967          -11237              20797

1304   1   4  1               376            -101               -109
   8   d   o  1              -268              71                 77
 163 -25   3  4              3315             752                 68
       -2   0 -1             2358             522                 32

1304 163   3  1            384680          118501              26406
1304   n   o  3            -21370           -6562              -1476
 163 -25   3  4            -77425          -25428               5216
       -2   2 -3             8004            1346                498

1305   2   6  1                46             -82                 10
 145   d   o  4                 8              -2                  4
   9  -3   3  1             -6958          -14500              -6380
       -2  -1  0             1160             772                760

1308   1   3  1               -56               9                -12
  12   d   o  1                37              -7                  4
 109   2  12  1               -56               9                -12
        0   0  0                37              -7                  4

1308 109   3  7             61694           -8175             -20928
1308   d  r1  2             -2619             677               1276
 109   2  12  1              -218            -981                  0
        0   0 -1              -235             -15                -36

1309  14   5  1             58219          -32305              40306
1309   d   o  2             -1609             893              -1114
   7  -1   3  1         -224028791       127132698         -147170870
        2   1 -1           6100383        -3307898             4438894

1311   2   5  1               -42             -34                -49
  69   d   o  1                -2               0                  5
  19  -7   3  1         -393802525        -224279118         -52408674
       -2  -1 -1          -46247609         -27170910          -5709978

1313   2   6  1             20622          -15736               6137
 101   d   o  1             -1986            1604               -585
  13   5   3  1          541663709        -425410586         160369416
       -1  -1  0          -53927271          42312206         -15969904

1313  26   6  1            -19461          -11557              -8190
1313   d   o  4               537             319                226
  13   5   3  1          -1079260          849511            -315120
       -1  -1 -2            -29784           23317              -8942

1316   1   5  1                43             -30                 37
 188   d   o  1                -8               3                 -6
   7  -1   3  1           27446497        -15282050          19006330
       -1   1 -1           -4014358          2220450          -2776160
```

```
1317  878  3  1         854294           -34242            94385
1317   n   o  1         -23528             942            -2603
 439  -28 18  1        1290221          -153211          -218622
       0  0 -1          33333            -4485            -6144

1321 2642  6  1     -91231672143       4741420386      -105633765
1321   n   o  1      2510118509       -130453896         2906373
1321  71   9  1    -4034448389353     28520819325      356933495907
       0 -1 -1     110978730873       -786683343       -9822345711

1324  331  5  1      18894049651      -1766698267       1825313071
1324   n   o  1     -1038510891         97106519       -100328280
 331  -1  21  1      -357835163         29783380       -119870326
       1 -1 -1        50196538          -4901834         161092

1332   3   3  1          -165               9               36
 12    d   o  1          -120               3               18
333   -30 12  3          -165               9               36
       0   0  0          -120               3               18

1332   3   5  1          -165               9               27
 12    d   o  1           102              -6              -15
333   -3  21  3         -77757            7722            -6606
       2   1  0         -43188            4296            -4146

1332   3   3  1           2517            2874             1872
444    d   o  2           -240            -272             -178
  9   -3   3  1         -29157           17874           -12222
       0   0 -1          -2388            2130             -876

1332  111  3  1         -73815          -13431           -10656
444    d   o  2           7104            1271              994
333   -30 12  3       -2069817         -367299          -292707
       0   0 -1         193362           35343            27579

1332  111  5  7          49173            9435             4995
444    d  r2  2          -4662            -896             -475
333   -3  21  3            111               0             -666
       2   1 -1           -444              42               24

1333   62  5  1          -1085           -1147                0
1333   n   o  1            31               31                0
 31   -4   6  1        31653480         17892859         5330667
      -2  -1 -1        -1074770          -441719         -220565

1333   86  5  1         -37926           14921            -6407
1333   n   o  1           1040            -409              175
 43    8   6  1       -43459713        -17318336        -11950345
      -2  -1 -1         1186407          475102           330141

1333 2666  3  1      84078997661       7034444949       741762513
1333   n   o  1     -2302685049       -192658307        -20317695
1333  -70 12 12          3999             2666             1333
      -2  -4 -1          -883              -12               29
```

1333	2666	5	1	1104858522965	-36599957056	-107574401008
1333	n	o	1	-30196006365	1000243546	2939978056
1333	38	36	3	13954239813022	-867665945271	447452995445
	-2	-1	-1	-382214849172	23765283603	-12254354391
1337	14	5	1	111524	-239099	-435050
1337	n	o	1	-3050	6539	11898
7	-1	3	1	81542307	-183392279	-330574587
	1	2	-1	-2234921	5017217	9037619
1339	2	4	1	-113841	-32950	-9870
13	d	o	1	31377	9150	2670
103	-13	9	1	-113841	-32950	-9870
	0	0	0	31377	9150	2670
1339	26	6	1	19721	1586	39
13	d	o	1	-5447	-442	-13
1339	-73	3	9	-47463	-6760	-1690
	-2	-1	0	22919	1040	-390
1339	26	6	1	275574	24869	13221
13	d	o	1	118908	10731	5705
1339	8	42	9	-3796	169	533
	-1	1	0	1658	-79	-121
1340	67	3	1	-1206	804	-134
1340	d	o	2	57	-42	12
67	5	9	1	7035	-5762	-9246
	0	0	-1	1218	-30	-336
1341	2	6	1	-5	206	205
149	d	o	1	-17	-10	3
9	-3	3	1	-14104069861	-16206670698	-10674646974
	-1	0	-1	1173429087	1320601870	854056562
1343	2	6	4	-212	68	0
17	d	o	1	42	-16	4
79	17	3	1	12442	-476	-5168
	0	0	-1	3112	-156	-1252
		c	-3	40	-10	-8
				0	2	-2
1349	38	5	1	87324	50540	11191
1349	n	o	1	-2378	-1376	-305
19	-7	3	1	113970303	124871534	-16894876
	-1	1	-1	-7526829	-2748522	-1824164
1351	2	6	1	-21	18	4
193	d	o	1	1	0	2
7	-1	3	1	-191	-193	-965
	1	1	0	39	-43	-33
1351	386	6	1	81934290	7205462	1941194
193	d	o	1	-5948132	-519742	-138390
1351	-52	30	3	382299418	-26945116	-36899284
	0	-1	0	27668368	-1942720	-2646384

1351	386	6	1	51917	−1737	−4825
193	d	o	1	−3499	145	359
1351	29	39	3	34740	−3088	−5597
	−1	−1	0	−4586	42	289
1355	2	6	1	114	−19	6
5	d	o	1	−70	11	−2
271	29	9	1	19777	−1420	−5370
	1	1	0	11709	−132	−1986
1359	6	5	1	−69933	51774	−27642
453	n	o	1	3279	−2440	1294
9	−3	3	1	−29507691387999	21900693094110	−11650329747036
	−2	−1	−1	1386197996331	−1028907592158	547601175972
1359	906	5	1	−25950113607	−2242598244	−292008330
453	n	o	1	1219242705	105366454	13719746
1359	69	15	3	−20233906799025	−1722038992740	−245070402498
	2	1	−1	925112425605	81196061964	9593187162
1359	906	5	7	140230597101	−5957389863	−13157804931
453	n	r2	1	−6588223167	279871579	618155867
1359	−12	42	3	906	−1359	0
	−2	−1	−1	−906	−15	−42
1364	1	3	1	−43	22	33
44	d	o	1	10	−6	−11
31	−4	6	1	1090	−561	−946
	0	0	−1	−297	181	292
1365	2	3	1	−44	42	−21
105	d	o	2	−6	4	−1
13	5	3	1	−44	42	−21
	0	0	0	−6	4	−1
1365	14	5	1	−491960	33208	−127638
105	d	o	2	−48182	3300	−12382
91	−16	6	3	277457139554	−95746371000	−119960875860
	−3	−3	0	−27079205760	9344022808	11706431672
1365	14	3	1	56	−35	0
105	d	o	2	12	−1	2
91	11	9	3	56	−35	0
	0	0	0	12	−1	2
1365	14	3	1	−1379	3255	5880
1365	d	o	4	39	−89	−158
7	−1	3	1	−18571	39690	72870
	0	0	−2	473	−1118	−1978
1365	26	3	1	47203	−19929	−72618
1365	d	o	4	−1281	537	1964
13	5	3	1	−1149577	−759486	−627354
	0	0	−2	−34451	−19154	−11854

```
1365   182   5   1     -916825      59696      -235326
1365    d    o   4       24855      -1604         6372
  91   -16   6   3    17077697   -1112475      4398030
      -3   -3  -2     -464551      30011      -118724

1365   182   3   1     1083173    -273455       128310
1365    d    o   4      -29275       7395        -3492
  91    11   9   3    29706677   -3900260    -13018460
        0    0  -2      853887    -118108      -346436

1368     1   6   4         -91        -24           -4
   8     d   o   1          70         16            2
 171    24   6   3       -2279       -420         -168
         0   0  -2         912        322          -28
             c4            -7          2            4
                            0          1            2

1368     1   6   1         262         69           37
   8     d   o   1        -184        -49          -26
 171    -3  15   3      564073     -79692        69924
        -2  -1  -2      399912     -56392        49258

1368     3   3   1         255        -36          -18
  24     d   o   1          24        -12          -24
 171    24   6   3         255        -36          -18
         0   0   0          24        -12          -24

1368     3   5   1        -768         27          117
  24     d   o   1        -138         45           72
 171    -3  15   3    14739771   -1820340     -3925152
        -2  -1   0     6066600    -750042     -1596366

1368     1   3   1         443        448          320
 152     d   o   1         -62        -80          -48
   9    -3   3   1       75265     -56064        29952
         0   0  -1       12288      -9088         4816

1368    19   3   1        5035        988           38
 152     d   o   1        -646       -196          -48
 171    24   6   3      749455     181602        25764
         0   0  -1     -119928     -29776        -4508

1368    19   5   1       -3952      -1045         -551
 152     d   o   1         646        169           90
 171    -3  15   3     2971771    -435708       726180
        -2  -1  -1      782040    -107744        38402

1368     3   3   1         267        252          180
 456     d   o   2         -24        -24          -18
   9    -3   3   1        2811       2808         1620
         0   0  -1        -216       -282         -204

1368    57   5   4       19551      -4104        -4788
 456     d   o   2       -1824        384          450
 171    24   6   3       78033     -18468       -20520
         0   0  -1       -8208       1518         1884
             c4          25593      -5358        -6270
                         -2394        501          588
```

1368	57	5	1	−16302	−2565	−513
456	d	o	2	−684	−345	−264
171	−3	15	3	−148843161951	−39429218988	−21009720060
	−2	−1	−1	−13953661560	−3691018392	−1969375854
1379	2	4	3	170	−1540	476
197	d	r	1	164	12	88
7	−1	3	1	−63	46	−36
	0	0	0	−5	2	−4
1381	2762	6	7	40080763	−1466622	1722107
1381	n	r1	1	−1078551	39466	−46341
1381	−31	39	1	−658737	−58002	−19334
	1	0	−1	−17779	−1582	−658
1385	2	4	1	183	−14	−28
5	d	o	1	93	−8	−12
277	26	12	4	−2082	−405	−295
	0	2	−1	1024	167	135
1385	554	4	7	−7869016	−3627592	−1659784
1385	d	r2	2	567434	41192	61712
277	26	12	4	968946	−116340	−44320
	0	2	−1	7592	−2192	2808
1387	2	6	1	13	−4	3
73	d	o	1	−1	0	−1
19	−7	3	1	2	−73	−146
	−1	0	0	18	−11	−8
1387	146	6	1	21900	1752	1387
73	d	o	1	2390	206	175
1387	65	21	3	5840	584	511
	−1	−1	0	888	62	47
1387	146	6	1	2086561336	185334006	81062120
73	d	o	1	244207450	21692042	9488200
1387	−16	42	3	53740113182	−3016008140	−5303911944
	−1	0	0	−6401652032	343062344	616430504
1389	926	5	1	21765630	2438158	1980714
1389	n	o	1	−584010	−65420	−53146
463	23	21	1	−94932254158	−14285498304	−9811123716
	1	2	−1	2675183244	376004228	241475128
1393	14	5	1	−959097328	535178917	−664093983
1393	n	o	5	25697278	−14339151	17793197
7	−1	3	1	−1532286	3074351	5170816
	2	1	−5	−29376	75885	146626
1393	398	5	1	−28813159653	3033196407	7873117993
1393	n	o	5	771996493	−81269011	−210945955
199	11	15	1	−559588	51541	208950
	2	1	−5	23054	−2621	−4832

```
1393 2786  5  1             -42496411195              3473517936              3732675835
1393    n  o  5              1138614465               -93066630              -100010297
1393  -73  9  3          210296067186086           -1336351569679          15807135350244
       -1 -2 -5            -5634556905414             35801623419            -423523422438

1393 2786  5  1           911915031499724          -17162777555854          -83099190676852
1393    n  o  5            -24433113712622           459845579022            226492496724
1393   62 24  3  30060743070535269712 2 -213353478557 40204692 55526890565580609308
       -2 -1 -5  -8054688789507572616  57160393174 3345040 -14880450655144 5432

1395    2  6  1                 -2791                    -490                     -17
   5    d  o  1                  1185                     230                      19
 279   33  3  3              -5491093                  431490                  729000
        2  1  0              -1149735                  434254                  340112

1395    2  4  1                    94                      12                      12
   5    d  o  1                    24                       8                       4
 279  -21 15  3                    94                      12                      12
        0  0  0                    24                       8                       4

1395    6  5  1                  -603                    -804                    -612
 465    d  o  2                    27                      38                      28
   9   -3  3  1              10698261                -8184465                 5301000
       -1  1 -1                561255                 -405267                  171762

1395  186  5  1                -11718                    1860                    2139
 465    d  o  2                   558                     -86                     -97
 279   33  3  3           -1150446549               214793730               220542525
        2  1 -1              57139665                -9260760               -10186539

1395  186  3  1                 66123                   -9486                    3999
 465    d  o  2                 -3069                     440                    -185
 279  -21 15  3                -60357                    4185                   10881
        0  0 -1                 -2325                     117                     531

1397  254  5  1                191389                  -48133                   10160
1397    n  o  1                   549                    -155                      24
 127   20  6  1              -4193540                  241681                 1370457
        2  1 -1                113490                   -5961                  -36347

1407    2  3  1                 -6141                    1246                    3108
  21    d  o  1                 -1271                     290                     692
  67    5  9  1                 -6141                    1246                    3108
        0  0  0                 -1271                     290                     692

1407   14  5  1                  1568                    -252                    -182
  21    d  o  1                  -186                      54                      60
 469  -43  3  3             -24652390                  158760                -3182004
       -1 -2  0              -5358396                   31356                 -697752

1407   14  5  1                187894                   -5313                  -29939
  21    d  o  1                -41284                    1163                    6571
 469   38 12  3                  -406                      63                     147
       -2 -1  0                   220                       5                     -17
```

```
1407    2  5  1              723              71              383
 201    d  o  1              -51              -5              -27
   7   -1  3  1            46433           52059            37989
        2  1 -1            -3961           -2131               97

1407  134  5  1            -7772           -1608              335
 201    d  o  1              544             114              -23
 469  -43  3  3         20331083        -3038517         -3182433
       -1 -2 -1          1440655         -213883          -224033

1407  134  5  1         53290862        -1506294         -8484478
 201    d  o  1          3756992         -105874          -598462
 469   38 12  3  383284376916787255610 -10817446575993772848 -61038723092040618876
       -2 -1 -1  27034795536311195816  -763003856637536608  -4305339571369298120

1413    2  6  1             2559           -1115            -2997
 157    d  o  1              211             -75             -233
   9   -3  3  1        -31082701        12310370         35387800
       -1  0  0         -2484211          978310          2821600

1413  314  6  4          -299556          -24178           -23550
 157    d  o  1            23864            1932             1880
1413  -75  3 12             -314           -1256             -628
        2  0  0             1256               4               52
            c-2             4082             314              314
                           -314             -26              -24

1413  314  6  1          1105437           83367            22451
 157    d  o  1            69551            7729             3493
1413   33 39  3      19469875679      -1122299428      -1774539914
        0 -1  0      -1554931297         89475236        141587382

1415    2  6  1              859            -154               18
   5    d  o  1              379             -68                8
 283   32  6  1        -20733743        -2303440          -914935
       -1 -1 -1         -3736063        -1158190         -1528869

1417    2  6  1             -346              61              -55
  13    d  o  1              -96              17              -15
 109    2 12  1          2349869         -443430          -935285
        0 -1 -1          -656317          121720           258715

1417   26  6  1         11070878         1024322           857623
  13    d  o  1         -3594980         -273002          -190029
1417   59 27  3  245206235670386355781  20494199505288583594  15741005630816883664
        1  0 -1  -68007974421770884269  -5684068224462438066  -4365769373983377240

1417   26  6  1             1196             117               39
  13    d  o  1             -326             -31              -11
1417  -49 33  3        -30066387         2059954          3032640
       -1  0 -1          9039725         -589390          -797224

1417    2  4  1             -769             393             -236
 109    d  o  1               45             -55               10
  13    5  3  1           184341         -137278            55374
        0  0 -1           -16635           13770            -4850
```

```
1417  218   6   1           135093728              11285642              8664301
 109    d   o   1           -12929726             -1081178              -830793
1417   59  27   3  1571158403918705921   13135220212336234   10088780650017676
        1   0  -1 -15053056556423223    -1258125127664718    -966330166338588

1417  218   6   1               -9919                  -327                 -436
 109    d   o   1                  63                   -47                   18
1417  -49  33   3           -515333579             150011686             490578044
       -1   0  -1           -589076389              28272618              13198108

1417   26   4   1               35347                -31733                26689
1417    d   o   2                -939                   843                 -709
  13    5   3   1                 286                  -611                   26
        0   0  -1                 -22                     7                   -8

1417  218   6   1            81201730             -14326306              11419276
1417    d   o   2            -2157252               380554               -303372
 109    2  12   1 -933858516686235286  -264831692982403092  -1398177691350931400... 
        0  -1  -1 -24794345924156200   -7037787578210480    -3712112518291352

1417 2834   6   1         107259371479            8964685925            6885523242
1417    d   o   2          -2849379679            -238149763            -182916138
1417   59  27   3       -243129078371036        17004593096686         -4943008283759
        1   0  -1         -6572712931752          442211878692          -138625105155

1417 2834   6   1             -1309308               123279                 51012
1417    d   o   2                34828                -3271                 -1354
1417  -49  33   3           1201054868             -32583915              75752820
       -1   0  -1            -32688948               798495              -2031810

1419    2   5   4             -2510990             -1003992              -696036
  33    d   o   1              437640               174664               120808
  43    8   6   1          -14903753434          -5983968276           -4126816056
        0   0  -1           2607922136           1036508320             720681176
               c2              308536               123022                85304
                              -53606               -21454               -14836

1428    1   5   3                 -193                   -9                 -104
 204    d   r   2                   -3                   18                   -6
   7   -1   3   1                  409                 -306                  306
        3   3   0                   74                  -30                   48

1429 2858   6   1           -601994830             43928889             -7047828
1429    n   o   5            15924900             -1162075               186440
1429   71  15   1         8712856897433          696734314750          608218410800
        0  -1  -5          -230497590805         -18430282510          -16089564280

1435    2   3   1                  202                  220                   60
 205    d   o   2                   20                   12                    8
   7   -1   3   1                  202                  220                   60
        0   0   0                   20                   12                    8

1448    1   6   1                 5575                 -598                  813
   8    d   o   1                -3940                  423                 -575
 181   -7  15   1           265347113             -25182316              61650636
        1   1  -1          -260975034              30334230             -21850588
```

1448	181	6	1	651419	−69866	95025
1448	d	o	2	−34226	3675	−4993
181	−7	15	1	12283136573	−1317331756	1797575436
	1	1	−1	−646935186	69466470	−94080052
1449	6	5	4	−40290	2070	−13869
69	d	o	1	4758	−216	1707
63	15	3	3	15305793	−5428782	−6243534
	0	0	−1	1882251	−655458	−737706
			c3	−10728	3735	4188
				−1260	447	516
1449	6	5	1	93	48	42
69	d	o	1	−9	−6	−6
63	−12	6	3	−69546	22977	−6831
	1	2	−1	−8694	2625	−933
1449	2	5	1	478	536	341
161	d	o	1	−38	−42	−27
9	−3	3	1	−31542313	23426466	−12452223
	−2	−1	−1	−2486967	1844720	−982559
1449	14	5	4	−23359	1127	−8211
161	d	o	1	1841	−89	647
63	15	3	3	−3367	5313	11592
	0	0	−1	−2415	523	158
			c2	693	−98	84
				−21	−4	−20
1449	14	5	1	−3025246	305144	1258278
161	d	o	1	236600	−24812	−99742
63	−12	6	3	1258956461172850470	52727146267991580	39771864716812536
	1	2	−1	9923869756753008	4154880166549904	3134657765584840
1453	2906	6	1	471297986	−34847299	−42295377
1453	n	o	1	−12363654	914183	1109615
1453	−67	21	1	2247187460125	−166371074716	−202227891962
	−1	−1	−1	−59119644691	4367056732	5293810758
1457	62	3	1	−185977369814	43279943944	−66735031884
1457	n	o	1	4872258536	−1133853416	1748332768
31	−4	6	1	1387939766306	−776765404244	−1280511938568
	0	0	−1	36360343696	−20349603624	−33547385912
1461	974	5	1	−2419975563	−348611645	−112862737
1461	n	o	1	63310951	9120511	2952637
487	−25	21	1	1575139233641	−98579721462	167122203354
	−1	−2	−1	47044229795	−1738556746	4644445102
1463	2	3	4	−1003	924	1232
77	d	r	1	−93	116	144
19	−7	3	1	17	−10	−18
	0	0	0	1	−2	−2
1463	14	5	1	−34377	−8505	−6335
77	d	o	1	3917	969	723
133	17	9	3	−828275	−201894	−147994
	1	2	0	92301	23178	17566

```
1463   14   5   1          -413          -112           -42
  77    d   o   1          -183           -48           -18
 133  -10  12   3         -2065             0         -1386
      -1  -2   0           663           -96             6

1463    2   5   1           750         -1635         -2892
 209    d   o   1           -52           113           200
   7   -1   3   1     -173399983     -139045401     -61894305
       1   2  -1      -11993049       -9618411       -4279677

1463   38   5   1        106286         26239         19589
 209    d   o   1         -7352         -1815         -1355
 133   17   9   3        184585         46607         34485
       1   2  -1        -13065         -3167         -2379

1463   38   5   1       3039620       -1045304        672866
 209    d   o   1       -198050         69580        -50866
 133  -10  12   3
      -1  -2  -1

1464    1   5   4           -55             0           -12
  24    d   o   1             6            -6             2
  61   -1   9   1          3421          -780           792
       0   0  -1         -1478           292          -334
          c4               5            -4            -3
                          -1            -1            -2

1464   61   5   4        -17995          2928        -12444
1464    d   o   2           938          -154           650
  61   -1   9   1         24217        -24156          7320
       0   0  -1         -4718             8          -986
          c4             549           244            61
                          45             9             4

1467    6   3   1        -45642        -31224         -8580
 489    n   o   1          2064          1412           388
   9   -3   3   1          1302          -828           -72
       0   0  -1            24           -24            36

1467  978   5   1    -5219532210     -458175396    -130817280
 489    n   o   1     236035410       20719408       5915762
1467   51  33   3   -23827928430     1516097556     2107873620
       2   1  -1    -1090387848       67431780       94999368

1467  978   5   1   -41372714442     1478745780     3079674078
 489    n   o   1    1072793628     -138651068     -168529474
1467   24  42   3 -283345730309216838 -25482207616224012 -10388070933382008
       2   1  -1  -12813321916640352 -1152346056726480  -469766151653328

1468  367   5   1     -239522183       38716298       -5178370
1468    n   o   1       12497259       -2021840         269573
 367   35   9   1     9593319445     -1446134874     243243196
       1   2  -1     -508968468       74241582       -13756014

1469    2   6   1           -36           -14            -4
 113    d   o   1            -2            -2            -2
  13    5   3   1          4070          2712          2260
       0  -1   0           428           236           144
```

```
1469   26   5   1          -2844985              2240407              -841685
1469    d   o   2            74495               -58293               22077
  13    5   3   1   -932970188828552771    732379793604395612   -276261272054638546
        0  -1  -1   2434220434178 3843    -19108521351571716     7207644670308170

1476    3   5   1              207                 -459                 156
 492    d   o   2               60                  -17                  17
   9   -3   3   1         -42428800797          30837023730          -13867869438
       -2  -1  -1          -3485677812           2646204966          -1637055504

1477   14   3   1            -20797               29288                58989
1477    n   o   1              541                 -762                -1535
   7   -1 . 3   1            20811               -39718               -64736
        0   0  -1             -327                  918                 1832

1477  422   5   1           170291348           -29229619           -46014036
1477    n   o   1            -4431006             760559             1197292
 211  -13  15   1          -8074441023         -1717990274          -627403014
       -1   1  -1           -210027235           -44707066           -16308694

1477 2954   5   7         5831077261016        -49388295250        -492004443343
1477    n  r2   1         -151724762990         1285084566          12801966809
1477   74  12   3             2954                -1477                1477
        1  -1  -1             -960                   43                  31

1477 2954   5   1       -23123779461405      -1920237223639       -435523508427
1477    n   o   1        601684149899          49964855591          11332385883
1477  -61  27   3   -80857815581738690887   1556410859410519210  -5278220441068780436
       -1   1  -1   -2099564260139582691    40192182698018442   -137735650751308428

1480    1   6   3                9                  -4                  -3
  40    d   o   2                0                   1                  -1
  37   11   3   1             1161                 -140                -820
       -1  -2   0              398                  -36                -250
               b2               43                   20                  20
                               -17                   -6                  -4

1480   37   6   1            209013             -105968               19869
1480    d   o   4            -10866                5509               -1033
  37   11   3   1          -42395303             4331960            28435980
       -1  -2  -2          -2233622              240354             1475540

1484    7   5   1             -2331                5250                9394
1484    d   o   2              123                 -271                -487
   7   -1   3   1         -20083477295         -16105754798         -7167988604
        1   2  -1          -1042679170           -836160840          -372113426

1489 2978   6   1        -6078770542586       -186262750368        -319297488126
1489    n   o   3        157531860198          4827015158          8274621802
1489   77   3  19             2978                -5956                5956
       -5  -3  -3             3868                 -160                -148

1491    2   3   1             -220                  471                 855
 213    d   o   1               14                  -33                 -59
   7   -1   3   1            14465               10266                3240
        0   0  -1              931                  838                 464
```

1497	998	3	1	3781285835874	-426473170348	152906160828
1497	n	o	1	-97730159536	11022517944	-3951974048
499	32	18	1	598942669090	-66698060552	21801124372
	0	0	-1	-14855359904	1697532808	-663228056
1501	38	5	1	-2755000	-1595658	-355699
1501	n	o	1	71110	41186	9181
19	-7	3	1	-3261527365	480377038	-1683905856
	-1	1	-1	-84186253	12389790	-43471496
1501	158	5	1	-1473113	402189	103095
1501	n	o	1	38023	-10381	-2661
79	17	3	1	-710247525	-209665684	-184962226
	-1	1	-1	-18622325	-5310524	-4786458
1501	3002	5	1	-5199315401	-413136741	-50936435
1501	n	o	1	134201023	10663591	1314735
1501	-73	15	3	-20002328961473	1511330215222	1720917877242
	-1	-2	-1	513562113705	-38982237574	-44610855346
1501	3002	5	1	176212137494	14662101222	4514925445
1501	n	o	1	-4549299768	-378544278	-116570845
1501	-46	36	3	-484515029323	33072011819	55075427490
	-1	1	-1	-16100531139	951149013	1208244816
1503	6	5	1	279	312	-1845
501	n	o	1	-9	-10	85
9	-3	3	1	-13058729823	-15548757444	-9821699190
	-1	1	-1	-616650339	-670003548	-451924446
1505	2	6	1	222	-57	-59
5	d	o	1	-148	17	25
301	-31	9	3	-62933253	11451650	13514270
	-1	-2	-2	29208801	-4936474	-6013118
1505	2	4	1	231	-4	-85
5	d	o	1	223	-18	-31
301	23	15	3	-150815	8336	30746
	0	0	-2	-64403	4264	14130
1505	14	3	1	24171	-55125	-100205
1505	d	o	2	-623	1421	2583
7	-1	3	1	35049	28665	12985
	0	0	-1	917	721	315
1505	86	5	1	-435687352	166525928	-74146190
1505	d	o	2	11239238	-4294280	1905582
43	8	6	1	-15383518332874 62968094	9332819989886905 3860	6781431373418055 18040
	-2	-1	-1	-20759995303161733432	9953348905537096528	2270693503544894 7256
1505	602	5	1	-14930202	-2592513	-429226
1505	d	o	2	384856	66827	11064
301	-31	9	3	7624932	1348480	233275
	-1	-2	-1	-200320	-34178	-5417

1505	602	3	1	−2396261	−433741	−320565
1505	d	o	2	61765	11181	8263
301	23	15	3	1383095	−201369	90601
	0	0	−1	38091	−5297	1879
1516	379	5	1	11083433552	−1460869523	489421271
1516	n	o	1	−569317740	75039827	−25139882
379	29	15	1	−11819432981	−394906630	−1013581682
	−2	−1	−1	−91207758	−88306336	−29282638
1517	2	6	1	−173	86	−20
41	d	o	1	27	−14	2
37	11	3	1	18780	−9676	2009
	−1	−1	0	−3036	1518	−253
1517	74	5	1	−182373	92685	−17427
1517	d	o	2	4695	−2379	443
37	11	3	1	−986700412159	107836281774	648621372024
	−1	−1	−1	−24782343945	2489052858	16705547040
1524	1	3	1	−97	−24	−21
12	d	o	1	58	14	11
127	20	6	1	−1334	−603	−354
	0	0	−1	1411	185	238
1529	278	5	1	−37127873	−7963171	−7162392
1529	n	o	1	949503	203649	183170
139	23	3	1	277646801	−72919539	6640447
	2	1	−1	7123397	−1864927	164037
1533	2	3	4	2147	−315	504
21	d	r	1	−457	67	−114
73	−7	9	1	67	−40	−118
	0	0	−1	63	−16	−14
1533	14	5	1	367500	46683	42686
21	d	o	1	83942	10663	9750
511	44	6	3	579488	84315	87381
	1	2	−1	159938	17981	14197
1533	14	5	1	5299	154	595
21	d	o	1	1373	6	95
511	−37	15	3	377795516616077	−47656348954494	−60410215779354
	−2	−1	−1	−82979866574095	10325529241658	13166807284102
1533	14	5	1	−359625	199675	−248906
1533	d	o	2	9189	−5097	6358
7	−1	3	1	−7643905676239	4486945601082	−4603557949326
	2	1	−1	187835661689	−97986117766	147511899706
1533	146	3	4	58619	−18396	−29127
1533	d	r	2	−1443	462	757
73	−7	9	1	6935	−1898	−3504
	0	0	−1	−169	62	88

1533	1022	5	1	2557582083	324885624	297068317
1533	d	o	2	-65322197	-8297776	-7587305
511	44	6	3	-2529278815	333024321	-30831696
	1	2	-1	-64719637	8488973	-803906
1533	1022	5	1	-1162525	31682	-127750
1533	d	o	2	29599	-822	3260
511	-37	15	3	23166603509	-2905620606	-3699070746
	-2	-1	-1	593377165	-74260798	-94288322
1535	2	6	1	-24328	3391	5311
5	d	o	1	-10906	1521	2385
307	-16	18	1	-68903	-12120	-4255
	-1	-1	-1	30153	5514	2055
1540	1	5	1	-355	793	1432
220	d	o	2	48	-107	-193
7	-1	3	1	-1579709	3560260	6410690
	0	3	-2	213908	-479206	-864130
1541	134	5	1	-878973	181436	598176
1541	n	o	1	22391	-4622	-15238
67	5	9	1	1143487272271	393603905854	232164822468
	-2	-1	-1	29134046965	10025491858	5915022356
1544	1	6	1	-165	27	-10
8	d	o	1	-98	23	-4
193	23	9	1	-195	88	4
	0	1	0	-312	26	-26
1544	193	5	1	15633	-3667	1930
1544	d	o	2	-808	185	-100
193	23	9	1	-1923662459	-398785092	-316172600
	0	1	-1	-98089918	-20288328	-16045418
1545	206	3	1	3513021	1018155	302305
1545	d	o	2	-89375	-25903	-7691
103	-13	9	1	220626	56135	21630
	0	0	-1	-4566	-1529	-324
1547	2	4	1	-1949434	101196	-486648
17	d	o	1	-439392	34632	-115992
91	-16	6	3	-1949434	101196	-486648
	0	0	0	-439392	34632	-115992
1547	2	6	1	389	116	79
17	d	o	1	95	28	19
91	11	9	3	1974	629	408
	-1	1	0	518	143	102
1547	2	3	1	66	-340	-816
221	d	o	2	20	-32	-44
7	-1	3	1	66	-340	-816
	0	0	0	20	-32	-44

1547	26	3	1	91	-65	-78
221	d	o	2	15	-3	-4
91	-16	6	3	91	-65	-78
	0	0	0	15	-3	-4
1547	26	5	1	7267	2171	1469
221	d	o	2	493	145	99
91	11	9	3	39585	20332	9282
	-1	1	0	4769	836	874
1548	3	5	1	33765	5295	5022
12	d	o	1	-19569	-3046	-2900
387	-39	3	3	-2102629467957	312168956292	-16339853754
	2	1	-1	-1213970306628	180228220962	-9436311606
1548	3	5	1	5514	963	372
12	d	o	1	-3186	-556	-215
387	15	21	3	-613005	-111258	-40662
	1	2	-1	374100	62874	25644
1548	1	5	3	59	-191	-4
172	d	o	1	33	-2	17
9	-3	3	1	23737	287670	65790
	-1	-2	0	-45924	-12470	-26740
			b2	-130	-430	-602
				-5	84	82
1548	43	5	9	43	-172	172
172	d	r	1	-387	21	18
387	-39	3	3	43	-86	86
	2	1	0	-172	14	12
1548	43	5	3	38872	6794	2623
172	d	o	1	-5934	-1037	-400
387	15	21	3	88795	-13674	-19092
	1	2	0	18060	-1300	-2606
			b1	-5590	602	1204
				-817	98	186
1549	3098	6	1	10205287543	-485642480	413355297
1549	n	o	1	256705221	-12240206	10393131
1549	11	45	1			
	0	1	-1			
1557	2	6	1	202	-354	35
173	d	o	1	-32	8	-15
9	-3	3	1	57220542515	-42489259488	22594720014
	-1	-1	-1	-4351903749	3228685944	-1718962338
1560	1	3	1	-11	12	-24
120	d	o	2	6	-4	-2
13	5	3	1	-11	12	-24
	0	0	0	6	-4	-2
1560	13	3	1	14183	8424	5772
1560	d	o	4	-724	-422	-294
13	5	3	1	-249119	192972	-60216
	0	0	-2	-11874	9452	-4202

```
1561   14   3   1        279079012681         223222849199          99627832087
1561    n   o   1         -7063593957          -5649853615          -2521617609
   7   -1   3   1               10353                 6713                 4067
        0   0  -1                -203                 -203                  -63

1561  446   5   4     -138242882144022        3786339327652      -24520785332260
1561    n   o   1       3498978936216         -95833661360         620630227368
 223  -28   6   1        6119863482          -1329809656          -1517035996
        0   0  -1         157168168           -33209688           -38338976
                 c3          -506656               -5352              -76266
                              12894                  144                1966

1561 3122   5   1  15069618953262657     1240336075088264      817672539897129
1561    n   o   1    -381417679351729       -31393368926158      -20695597121895
1561   41  39   9          32204991            -3894695              594741
        0  -3  -1           1557039             -37521               55297

1561
1561    n       1
1561  -13  45   9

1565    2   4   1                -623                   96                    9
   5    d   o   1                 211                  -42                    9
 313   35   3   7              131317                 -954               -24822
        0   0  -1               60831                 -798               -11082

1565  626   4   1           -40975769            -6408362            -6056863
1565    d   o   2             1035097              162108              153101
 313   35   3   7          4823144391          -801212392            63801294
        0   0  -1           116451185           -21108896             804130

1569 1046   3   1          90637743052         11302094852          1113354555
1569    n   o   1          -2288219124          -285330026           -28107487
 523  -43   9   1               95186               -1046                 9937
        0   0  -1                1990                 -48                  265

1577   38   5   4           193344969          -206522343          -265505297
1577    n   o   1            -4868745             5200573             6685861
  19   -7   3   1             2065908             1212713              280706
        0   0  -1               52584               29967                6356
                 c3             7752                4579                1159
                                196                 111                   21

1580   79   5   1             -505047             -148362             -131298
1580    d   o   2               25701                7365                6617
  79   17   3   1        53422820830519       -25918440867630      -196179710130
       -1   1  -1         3833232062322        -971812037328        286414128114

1581   62   5   1               -1550                1178                2635
1581    d   o   2                  40                 -30                 -67
  31   -4   6   1            -5633041             1294839            -2020518
       -1  -2  -1              140699              -33147               50508

1589   14   5   1            -3583741             1988658            -2479974
1589    n   o   1               89903              -49888               62214
   7   -1   3   1        -8462284834899        19013209903926       34259248350754
        2   1  -1         -212247445567          476950087630          859468780866
```

1591	2	6	1	21	1	-8
37	d	o	1	1	-1	-2
43	8	6	1	-5243675	-2072592	-1447514
	-1	-1	-1	852761	344288	236382
1591	74	6	1	217634	-2886	-17797
37	d	o	1	-35538	458	2929
1591	71	21	3	-2202783011630	635228614780	-896203345332
	-1	0	-1	-3276829009180	145510198608	96760688948
1591	74	6	1	1318330128	109271286	61421184
37	d	o	1	-216732180	-17964086	-10097582
1591	17	45	3	5821604124036482	-284483484407340	221626527689268
	-1	0	-1	957066096059652	-46768803607236	36435156417912
1592	1	6	1	-49	6	6
8	d	o	1	7	0	-7
199	11	15	1	-6271	980	-588
	1	1	0	-4494	672	-434
1592	199	5	7	-597	0	-1194
1592	n	ɍ		1	-243	43	24
199	11	15	1	-104077	15920	-11940
	2	1	-1	-5694	870	-440
1596	1	3	1	565	-48	-182
12	d	o	1	-316	26	106
133	17	9	3	31267	-1250	-8160
	0	0	-2	-13286	1900	5590
1596	1	3	1	19	-3	-3
12	d	o	1	-4	1	3
133	-10	12	3	454	-96	-141
	0	0	-2	-219	52	89
1596	7	5	1	-5747	-1631	-2800
1596	d	o	8	288	81	139
7	-1	3	1	755362069	625186716	297667566
	0	3	-4	-38453588	-29864498	-12318070
1596	19	3	1	78907	-11742	40470
1596	d	o	8	-3950	588	-2026
19	-7	3	1	55537	-8436	27702
	0	0	-4	-2710	386	-1448
1596	133	3	1	988855	-223440	76342
1596	d	o	8	-49504	11186	-3822
133	17	9	3	296191	-35378	-67032
	0	0	-4	7350	-84	-3934
1596	133	3	1	1740305	-187131	314013
1596	d	o	8	-87152	9361	-15723
133	-10	12	3	-19208924	2017344	-3499629
	0	0	-4	951777	-103762	174073

```
1597
1597    n    1
1597   50 36  1


1603    2  6  1            -848              -681              -304
 229    d  o  3              56                45                20
   7   -1  3  1           -13051            -10534            -4580
        0  1 -3             871               694               316

1603  458  6  3        -9488340658        -751103512        -593191524
 229    d  o  3          636688782          49455458          38375448
1603   65 27  3      -1879835434211      -6442126324       97703877294
       -1 -1 -3        -68062154223        3978015876        9904007454
           b2           -932029542          -73065885         -57194811
                          61569196           4828931           3780863

1603  458  6  1            1768338           144957             48548
 229    d  o  3            -116854            -9579             -3208
1603  -43 39  3             -5954            -25648             6412
        1  1 -3             30628             844               1272

1608    1  3  1             1567              208               304
  24    d  o  1             -188             -200               -44
  67    5  9  1             1567              208               304
        0  0  0             -188             -200               -44

1608   67  3  1            25661             8576              5360
1608    d  o  2            -1216             -444              -256
  67    5  9  1           1169619          -299088            216544
        0  0 -1            59776            -15172             10116

1609 3218  6  1  62260022632504767  -2299994038878016  3038526221645871
1609    n  o  1  -1552141290069335    57338811707918   -75750406282581
1609  -19 45  1 -23967607576070700 -1981190678051384  -853938066455789
       -1  0 -1    597216939880320    49401951032166    21274259079219

1611    6  3  1            -119574           88800            -47088
 537    n  o  1             5160             -3832             2032
   9   -3  3  1            -1434             1584             -2880
        0  0 -1             192              -120               -24

1612    1  3  1             4486             2714              1972
 124    d  o  1             -805             -488              -354
  13    5  3  1            35779            -25418             10864
        0  0 -1             5736             -4982             1650

1612   31  5  1            160673            1984             -11129
 124    d  o  1            -28872            -356              1997
 403  -37  9  3  -117353830129337   2529309740862  -15588518977094
        1 -1 -1    -21233210481078    477913014780   -2772190854186

1612   31  5  1            -1395             -124              -124
 124    d  o  1              -7               -18                -3
 403   17 21  3           -19851067          2186926          -1473988
       -1  1 -1            3840906           -411824           213690
```

1612	13	3	1	-6162	5460	-1768
1612	d	o	2	307	-272	88
13	5	3	1	-1547	832	2600
	0	0	-1	-88	28	124
1612	31	3	1	52421	-29016	-47151
1612	d	o	2	-2610	1446	2349
31	-4	6	1	192262	90675	36270
	0	0	-1	9369	4575	1770
1612	403	5	1	900507127	135432986	19404047
1612	d	o	2	-44853272	-6747032	-967325
403	-37	9	3	-485394101282361869	-72996539467030530	-10471229243816510
	1	-1	-1	24172399175729178	3636366522601980	520713146722710
1612	403	5	1	-2970513	-450554	-289354
1612	d	o	2	147963	22444	14415
403	17	21	3	394875321321	-27369889274	-73257003092
	-1	1	-1	19714537358	-1356455716	-3644840934
1615	2	6	1	-162	82	100
85	d	o	2	2	-12	-16
19	-7	3	1	-700058	746300	960160
	-1	1	0	75856	-81092	-104100
1621	3242	6	1	22344455665	-1681344967	-1811319989
1621	n	o	1	-554976897	41760441	44989025
1621	-79	9	1	-40274794634783	3034429021010	3273230075772
	-1	-1	-1	1001669204883	-75375075618	-81204951156
1624	1	4	1	47	-4	28
232	d	o	2	2	-4	2
7	-1	3	1	47	-4	28
	0	0	0	2	-4	2
1624	7	3	1	6167	17892	20916
1624	d	o	2	-306	-888	-1038
7	-1	3	1	1267	3024	11340
	0	0	-1	282	-342	-324
1628	1	5	4	23	0	-22
44	d	o	1	12	-2	-6
37	11	3	1	23	0	-22
	0	0	0	12	-2	-6
		c2		6	-1	-3
				1	0	-1
1628	37	5	4	-231139	117216	-21978
1628	d	o	2	11448	-5814	1086
37	11	3	1	2556811	-1355310	351648
	0	0	-1	-140946	68664	-7986
		c4		3737	-1073	-1258
				21	30	-78
1629	2	4	3	-61	54	-72
181	d	o	1	-9	6	0
9	-3	3	1	-61	54	-72
	0	0	0	-9	6	0
		b2		108	-41	-121
				-8	3	9

```
1629  362  6  3     -289753307     -22534138      -2156072
 181   d   r  1       21564159       1677046        160460
1629   78 12  3            362          -181           181
   0   -1  0             362            11            15

1629  362  6  3         869524         72581         52852
 181   d   o  1         -65160         -5363         -3940
1629  -57 33  3      181529063     -12938604       3524070
   0   -1  0          15487989       -796316        382946
            b1         -5903858        358199       -131587
                       -439106         26633         -9755

1631    2  6  3            -91           299           622
 233    d  o  1             13           -23           -36
   7   -1  3  1        -102052        239291        441302
   0    1  0            7462        -16113        -28372
            b2            9964        -22368        -40309
                         -652          1466          2641

1641 1094  5  4   -44962033594   -6202980000   -3044061564
1641   n   o  5     1109919944     153124996      75144836
 547   -1 27  4           7658         19692             0
   0    0 -5          -4540          -168          -324
            c4           43760         -3282          3282
                         1096           -78            84

1643    2  6  1           -103            28           -22
  53    d  o  1            -11             2            -6
  31   -4  6  1        2388076      -1338409      -2209729
  -1    0 -1         -329422        184103        303045

1645   14  5  1         -28238        -22610        -10059
1645    d  o  2            696           558           249
   7   -1  3  1    -6131209441   -4927488930   -2185573320
  -1    1 -1        151555881     121276026      54154512

1649    2  6  1             86          -100          -177
  17    d  o  1            124           -28           -15
  97  -19  3  1   -11088270024    4522421923    4780266720
   1    0 -1      -3081391940     984838985    1148410754

1649  194  6  1        -611876       -171690        -17460
1649    d  o  2          15068          4228           430
  97  -19  3  1      -23633274       5659368     -10019324
   1    0 -1        -1361764        -83396       -268560

1651    2  6  1            -17            -2            11
  13    d  o  1              7             0            -3
 127   20  6  1           -141           -26           -13
   0    1  0            -23            -8            -9

1651   26  6  1           6084          -143          -637
  13    d  o  1           2260             3          -139
1651   77 15  3          19201          -182         -1612
   0    1  0           5665           -54          -436

1651   26  6  1       15423174       -579358      -1376050
  13    d  o  1        4328748       -156532       -379226
1651   23 45  3   -81888348698    3018107092    7241277472
  -1   -1  0     -22719187440     837444564    2008104516
```

```
1652   1  5  1              -75                  -79                  -7
 236   d  o  1                9                   12                   4
   7  -1  3  1        -261777689           -211582850           -95816708
       1  2 -1          34221754             27228456             11902538

1653  38  5  1             90744               -28177               68039
1653   d  o  2             -3584                 -89                -1847
  19  -7  3  1 -12225720449476387   -860765922158358   -4889835612769014
      -2 -1 -1  120326962897967    -83159010913258     97121723382670

1655   2  6  3             -3051                 283                -354
   5   d  o  1             -1537                 145                -122
 331  -1 21  1           -456603               44400               -5810
       1  1  0            -92663                7928              -26082
            b1             -1986                 205                 395
                            820                 -91                -183

1656   1  5  1               -83                  38                  67
 184   d  o  1                -7                   2                  12
   9  -3  3  1            -20423                7912               23828
      -2 -1  0             -3128                1258                3468

1656   3  5  1            -10515              -11844               -7791
 552   d  o  2              885                1016                 658
   9  -3  3  1      -4309531299285      3359004825564       -1672334706168
      -2 -1 -1      -381377900136       269418855822        -153140349882

1657
1657   n     1
1657 -70 24  1

1659   2  5  1             -2892                1245                 -43
  21   d  o  1              -792                 225                 -51
  79  17  3  1          -2907385             -547722             -652050
      -1  1  0            432429              190026              133722

1659  14  5  1           2843848              387891              138957
  21   d  o  1           -619598              -84511              -30275
 553 -22 24  3              2954                 441                 147
       1 -1  0              -728                 -91                 -35

1659  14  3  1            -54250                4088               -3724
  21   d  o  1             11436                -932                 800
 553   5 27  3            -54250                4088               -3724
       0  0  0             11436                -932                 800

1659   2  5  1             32807              -18187               22766
 237   d  o  1             -2133                1185               -1472
   7  -1  3  1      3432033675989       -7807931144076       -14165627241486
       1  2 -1       230322459909        -511279613252        -915043772254

1659 158  5  1          -9259432            -1262973             -452433
 237   d  o  1           -589958              -80467              -28827
 553 -22 24  3         -230322367           11814450            -21264825
       1 -1 -1          14994105             -771440              1375505
```

1659	158	3	1	−20224	1343	−79
237	d	o	1	558	−31	125
553	5	27	3	3208743	−187704	−452986
	0	0	−1	176927	−16496	−31714
1661	302	5	1	−12630697	3101238	4085154
1661	n	o	1	309915	−76094	−100236
151	−19	9	1	296948899	81907232	12613634
	2	1	−1	8872251	1912792	621122
1665	2	4	1	26	2	3
5	d	o	1	−4	−2	−1
333	−30	12	3	446	−69	18
	0	0	−2	222	−29	8
1665	2	6	1	−529	−132	−85
5	d	o	1	−337	−50	−19
333	−3	21	3	−4830864853	−865440810	−427220070
	1	2	−2	−1996745145	−401829130	−222119198
1665	2	4	1	542	−380	220
185	d	o	2	−40	28	−16
9	−3	3	1	−1678	600	1860
	0	0	−2	−120	52	140
1665	74	4	1	489078062	−69455216	18617956
185	d	o	2	−35953936	5106304	−1369504
333	−30	12	3	1024270127762	−38951172948	−184577138544
	0	0	−2	75379407360	−2874194240	−13567579264
1665	74	6	1	98642	18796	9805
185	d	o	2	−7252	−1382	−721
333	−3	21	3	−2731081	−375735	−124320
	1	2	−2	125985	34385	23338
1668	1	5	1	98	30	37
12	d	o	1	−97	−16	−10
139	23	3	1	−594635712719	−134589600594	−123531841398
	2	1	−1	343313263166	77705240998	71321219504
1669	3338	6	1	−2347343353	157997554	203599641
1669	n	o	1	57341045	−3865462	−4990947
1669	−67	27	1	65208081669379333699	5629112686035031726	1462770565736870350
	0	−1	−1	178136901793794365	125459382104568978	20148760594663962
1672	1	3	1	−107	−16	−64
88	d	o	1	10	−4	12
19	−7	3	1	8561	−3824	5792
	0	0	−1	−3332	−56	−1428
1672	19	3	1	18867	−2812	9652
1672	d	o	2	−922	138	−472
19	−7	3	1	38095	−5776	18924
	0	0	−1	−1808	258	−970

```
1673    14  5   4        70846545        56830137        25280703
1673    n   o   1        -1732091        -1389411        -618075
   7    -1  3   1         978719          798021          346311
        0   0  -1         -24399          -19245          -8781
               c2           2366            1631             931
                             -48             -45             -15

1677     2  5   1           1727           -1344             490
 129     d  o   1            149            -118              46
  13     5  3   1        -1050058         825471         -317598
        -1 -2   0          -93074          72959          -26972

1677    86  5   1          -15910          -1247            -516
 129     d  o   1            482             115             170
 559    47  3   3         -253270         -40119          -48762
         1 -1   0           36224            3455            2404

1677    86  3   1            688             -43               0
 129     d  o   1             16              -1               6
 559    -7 27   3            688             -43               0
         0  0   0             16              -1               6

1677    26  5   1           5681           -2379           -8723
1677     d  o   4           -139              59             213
  13     5  3   1       -822268291      -380246334      -134911296
        -1 -2  -2        -14475107       -11642798       -11901688

1677    86  5   1         -172129          65747          -29025
1677     d  o   4           4191           -1603             717
  43     8  6   1       -487649713        61740432        265105191
        -2 -1  -2        -8834107         2735626          7324007

1677  1118  5   1          193973            5590          -51987
1677     d  o   4           11893            -230            -985
 559    47  3   3   -4550675369940193  25609194826254  616983548115792
         1 -1  -2   -111121009871047    625760710186   15066700280792

1677  1118  3   1          406393          -13975          -20124
1677     d  o   4           -5783             907             750
 559    -7 27   3        151033415        11368942        10191688
         0  0  -2          1461417          425938           72704

1683     6  3   1           -5187            3861           -2013
 561     d  o   2            219            -163              85
   9    -3  3   1           -291             297            -495
         0  0  -1             33             -21              -3

1685     2  4   1           -930             103             -79
   5     d  o   1            436             -41              39
 337     5 21   1         -56731            5926          -10646
         0  0  -1          42655           -4302            1174

1685   674  4   1         -126375           7751           -4718
1685     d  o   2          -1127             241            -252
 337     5 21   1        29621963        -2916398        -2364392
         0  0  -1         -96271            7134           112160
```

```
1687    2  6  1          1832            -590           -2313
 241    d  o  1          -118              38             149
   7   -1  3  1        -47716          202199          459346
        1  1 -1         10306          -17039          -24584

1687  482  6  1    -57410337970    -4360640504     -619855374
 241    d  o  1     -3707440820     -279294356      -38568642
1687  -76 18 39
        3  4 -1

1687  482  6  1       -794095          -65311          -46031
 241    d  o  1         51153            4207            2965
1687   59 33  3      -40185063          880855         3378579
        1  0 -1      -2584159           57869          218087

1688    1  6  1             5               1               4
   8    d  o  1            -6               1              -1
 211  -13 15  1      -91125971         7675692       -13348624
       -1  0 -1      64426920         -5425042         9442178

1688  211  5  1      -1475734          251512          397313
1688    n  o  1         71832          -12243          -19342
 211  -13 15  1    -19161092093     1613247764     -2809514796
       -1  1 -1      -933207142       78615516       -136636294

1691    2  6  1          -139             -78             -22
  89    d  o  1            15               8               2
  19   -7  3  1       5885928         3406742          755165
       -1 -1 -1       -624462         -360948          -80161

1692    1  5  1          -869            -832            -613
 188    d  o  1           103             139              80
   9   -3  3  1    -4791717984935    3556542773856   -1892399498358
       -1 -2 -1    -698944123884     518775433294     -276034261078

1693 3386  6  1      -3860040          455417          -57562
1693    n  o  1       -178300            4295           -5900
1693   47 39  1   -1936340499794439  109324410018350  -51939550724600
        1  1 -1    -47060363920585    2656989586890   -1262300301520

1705   62  5  1     -195080148       -91859138       -34949524
1705    d  o  8       4724550         2224618          846444
  31   -4  6  1  675444967741786502 -15636103544944762 0 245611396319342900
       -1  1 -4   16480320041661120  -385525614129674 4 5835271476273224

1708    1  5  1        -12808           -6169           -2103
  28    d  o  1         -6822           -1911           -1248
  61   -1  9  1   -143950420548395    30479076479502   -32908455141282
       -2 -1 -1    54291068940318    -11562597028896    12417720154998

1708    7  5  1          -994             133               0
  28    d  o  1           309             -52               6
 427   41  3  3     -82228398959      12201712978      -626836420
       -1  1 -1     31174222420      -4612518310       222065006
```

```
1708    7  3  1              -49            -14              -7
  28    d  o  1               40              2              -1
 427  -40  6  3            -2058           -231              14
        0  0 -1              615            113              24

1708    7  5  1             6531         -15106          -27069
1708    d  o  6             -316            731            1310
   7   -1  3  1            -4263           -854          -11956
       -1  1 -3             -390            372             166

1708   61  5  3          -469578          56669          -32635
1708    d  o  6            14256          -5823              96
  61   -1  9  1     225617249779    82141873002     39596212152
       -2 -1 -3      10928441616     3972142302      1910445510
            b1         -38464648      -13957776        -6740622
                        -1855623        -676698         -324870

1708  427  5  7         -1476566         219051          -11102
1708    d r1  6            71375         -10600             550
 427   41  3  3              427           -854             854
       -1  1 -3             -542             44              38

1708  427  3  1          -412909         -60634           -4697
1708    d  o  6            19984           2934             227
 427  -40  6  3             -854           -854            -427
        0  0 -3              281              6             -17

1713 1142  5  1    -328696281901    13289982175    -32160917779
1713    n  o  1      7941748141     -321104001       777051409
 571  -31 21  1       369378758      -40738566       -59110491
       -1 -2 -1         9404814       -1003640        -1381237

1719    6  5  1             2745           2892            1983
 573    n  o  1             -105           -128             -79
   9   -3  3  1    -113327316570    84116019528    -44755959636
       -2 -1 -1      -4734368952     3513918804     -1869768924

1720    1  4  1              -11             10              20
  40    d  o  2               12              0              -4
  43    8  6  1              -11             10              20
        0  0  0               12              0              -4

1720   43  3  1          1362283         544380          377110
1720    d  o  2           -65696         -26252          -18186
  43    8  6  1         -1354027         497510         -233060
        0  0 -1           -62826          25002          -10520

1729    2  6  1          5974157        1484922         1103389
  13    d  o  1         -1665093        -409998         -306655
 133   17  9  3 -1203441804013151706235 27214123643321196 0776 -92905577614219047522
        1  2 -2 -3337747024555532463851 75478398614397514416 -2576737107542400 1698

1729    2  6  1              395            -84            -136
  13    d  o  1             -107             24              38
 133  -10 12  3            15121          -4017           -5954
       -2 -1 -2            -4929            923            1580
```

```
1729   26   6   1             38610              -2795                  78
  13    d   o   1            -10626                781                 -16
1729   83   3   9        8760863189         -642183152            14161914
       -1  -2  -2       -2441629963          177287496            -4730246

1729   26   6   1             40014               -585                2080
  13    d   o   1              8960                -11                 748
1729  -79  15   9         731889665          -53176058           -61858056
        1   2  -2        -215629361           14857562            16321472

1729   26   6   1           -172939               3380               33566
  13    d   o   1           -134913               5302                6440
1729   29  45   9  50768010715654141    4021939301080818     2426529801808736
       -2  -1  -2 -14080902229470027   -1115471967107622     -672964769557992

1729   26   6   1           -384020              16224              -15795
  13    d   o   1            113426              -4792                4665
1729    2  48   9           -703703             -59761              -28496
       -2  -1  -2           -211025             -15905               -8562

1729    2   5   1               -22                 26                 -50
 133    d   o   1                 6                 -4                  -2
  13    5   3   1                 2               -532                2660
        1  -1   0              -252                152                 156

1729   14   5   1               385                 84                   0
 133    d   o   1                31                 12                   4
  91  -16   6   3              5467              -2128               -2926
       -1   1   0              -617                192                 218

1729   14   5   4           2901010            -734160              350588
 133    d   o   1           -253524              63876              -29640
  91   11   9   3           2901010            -734160              350588
        0   0   0           -253524              63876              -29640
               c2              4326              -1134                 728
                              -442                108                 -34

1729   38   5   1             12711               2413                  95
 133    d   o   1              1159                209                  19
 247  -31   3   3          -2741757             318402             -717668
       -1   1   0            492233              19190               65132

1729   38   5   1           3595123            -479484             -892791
 133    d   o   1            311931             -41538              -77397
 247   -4  18   3         -13407559            1747620             2878785
       -2  -1   0           -949617             129930              274665

1729  266   5   1            -57855               4522                -399
 133    d   o   1              5715               -394                 -17
1729   83   3   9          15677907             -25270            -1179710
       -1  -2   0           1370839              -2670             -102110

1729  266   5   1          -1228787              13965              -73283
 133    d   o   1            -97191                549               -7103
1729  -79  15   9          -9813139             129808             -542374
        1   2   0           -725313               2360              -57082
```

```
1729  266  5  1           -26387998              1324547              -871150
 133   d   o  1             2288106              -114851                75538
1729   29 45  9          -917850955            -73747436            -43986026
       -2 -1  0           -80974575             -6325092             -3859938

1729  266  5  1            51124801             -2159920              2100735
 133   d   o  1            -4434815               187362              -182231
1729    2 48  9             -248311                 7448               -10241
       -2 -1  0               17283                 -994                  707

1729   14  3  1             -916118              1714769              2728894
1729   d   o  2               22032               -41239               -65628
   7   -1  3  1               -3444                -2527                -1330
        0  0 -1                 -74                  -65                  -24

1729   26  5  1            -2662946              2593838              6929182
1729   d   o  2               64042               -62380              -166642
  13    5  3  1          -360676290            195819624            332223892
        1 -1 -1             2852476              -139436             -9713528

1729   38  3  1            14748161              8410597              1804335
1729   d   o  2             -354683              -202269               -43393
  19   -7  3  1                5472                 -247                 2470
        0  0 -1                  92                  -27                   56

1729  182  5  1          -139236552              5609968            -44134454
1729   d   o  2             3348462              -134940              1061398
  91  -16  6  3  331982742847650258  -21556885521983084  85180040763418328
       -1  1 -1    -7986668961840712     517608528671936  -2048686063146216

1729  182  5  4            79231607            -13588211            -20078877
1729   d   o  2            -1905465               326787               482883
  91   11  9  3            -2074618              -340613              -376922
        0  0 -1              -25842               -14265                -6204
            c3              -21203                -5915                -4277
                             -477                 -151                  -97

1729  266  5  1          -267667288             22251831             87727864
1729   d   o  2             6437212              -535141             -2109794
 133   17  9  3      2233402204208         561775813235         426244878696
        1  2 -1         55055844198          13397873841           9809078082

1729  266  5  1         -3047233490            329389130           -540975638
1729   d   o  2            73283664             -7921626             13010070
 133  -10 12  3
       -2 -1 -1

1729  494  5  1           473702034           -100402783           -106951247
1729   d   o  2           -11392204              2414617              2572103
 247  -31  3  3           -87307090            -16154047             -1061606
       -1  1 -1            -2111198              -385831               -22480

1729  494  5  1         50014495986          -6672202602         -12486311890
1729   d   o  2         -1202991888            160426410            300270702
 247   -4 18  3
       -2 -1 -1
```

```
1729 3458   5   7         -339779622           -18728528           -20983144
1729   d  r2   2            8171464              450408              504630
1729   83   3   9              3458               -6916                6916
       -1 →2 -1               4500                -172                -160

1729 3458   5   1       -135215089464          1169684061         -8927243689
1729   d   o   2         3251828726           -28130087           214693993
1729  -79  15   9        -116766286           -20341685           -57928416
        1   2  -1          19348694             -680597              70354

1729 3458   5   1          -99557549            -112385             5453266
1729   d   o   2            2398509               2491             -131008
1729   29  45   9  -827160088890348122   41518123286289195  -27298880199154564
       -2  -1  -1  -19892628317785224    998481580710447     -656519293223994

1729 3458   5   1        10821309133544        -457178694336       444655204994
1729   d   o   2         -260244947766         10994829180        -10693647982
1729    2  48   9 -224068812486140258 -17893924554243556  -9059137488087232
       -2  -1  -1   5379795226157344     430717957528832      218639762406432

1736    1   6   3                 67                  -7                  -6
   8    d   r   1                  0                   5                  -5
 217   29   3   3               -123                  40                 -20
       -1  -2   0                192                 -30                 -10

1736    1   6   3               -318                 -68                 -15
   8    d   r   1               -235                 -46                  -8
 217  -25   9   3                 37                  -4                   8
       -2  -1   0                -42                   0                  -6

1736    1   5   4                167                  84                  28
  56    d   o   1                 48                  22                  10
  31   -4   6   1                167                  84                  28
        0   0   0                 48                  22                  10
            c-3                   5                   6                   0
                                  4                   1                   1

1736    7   5   1              -2205                  77                 588
  56    d   o   1                706                   1                -137
 217   29   3   3             556703              108164              106428
       -1  -2   0             162370               28702               25360

1736    7   5   1               2436                1050                 497
  56    d   o   1               1373                 132                 -52
 217  -25   9   3             -47341              -13272               -4508
       -2  -1   0             -17376               -2574                   6

1736    1   5   7                898                -265                1139
 248    d  r1   1                134                -104                  27
   7   -1   3   1             -19343               43276               77500
        1   2  -1               2414               -5448               -9878

1736   31   5   1              48701                7347                5084
 248    d   o   1              -4066                -965               -1127
 217   29   3   3      -64014434660781      -11847218192552    -11067197458964
       -1  -2  -1      8126629877568       1505269630878       1405487987482
```

```
1736   31   5   1              -1860               1054                 961
 248    d   o   1                235               -134                -122
 217  -25   9   3             -885949             212412              304792
       -2  -1  -1             158910             -28908              -30786

1736    7   5   1               4116              -1855                2765
1736    d   o   4               -198                 90                -131
   7   -1   3   1          105563563         -297467072          -514507868
        1   2  -2            6964906          -12757050           -24019888

1736   31   3   1             -29915              16926               27776
1736    d   o   4               1434               -812               -1334
  31   -4   6   1             112871             -65534             -107198
        0   0  -2              -5674               3062                5088

1736  217   5   1          -54362189             816137            12352074
1736    d   o   4            2609436             -39169             -592919
 217   29   3   3     545374524714173      -8186839576440     -123919443680116
       -1  -2  -2    -26178932884920       392959745602       5948303533422

1736  217   5   1               -434                  0               -1953
1736    d   o   4                267                 48                 102
 217  -25   9   3         922935936349        -38282213844        157828601448
       -2  -1  -2      -44358109914         1849085076        -7561738362

1737    2   6   1                 48                -35                  27
 193    d   o   1                 -4                  3                  -1
   9   -3   3   1              -2893               2123                -965
       -1   0   0                193               -145                  89

1737  386   6   1             596370              26827                8106
 193    d   o   1              11966               2317                2980
1737  -75  21   3           -4804735            -395071             -348944
        0   1   0            -383105             -27983             -22218

1737  386   6   1        -10896375858         -929343636         -457147906
 193    d   o   1          811598968           65715912            30641050
1737    6  48   3      458357943464810     -19914124002064     -38162939429780
       -1  -1   0       33082943654032      -1426003455016      -2743460108656

1741 3482   6   1          116791503           -7070201          -11102357
1741    n   o   1           -2800025             169471             266029
1741  -49  39   1        -1072983523         -148852018         -811953652
        1   0  -1         -283360389           11424730            2463612

1743    2   3   1                -29                  9                 -15
 249    d   o   1                  1                 -1                   1
   7   -1   3   1                -29                  9                 -15
        0   0   0                  1                 -1                   1

1745    2   4   1              60932              -1050                7967
   5    d   o   1             -24300                -46              -4107
 349  -37   3   4                 57                520                1050
        0   2   0              -2821                256                  46
```

```
1745  698  4  1           2978017          -13960            433807
1745    d  o  4            -71289             334            -10385
 349  -37  3  4            220568           43625              6980
        0  2 -2             -6646            -813                70

1748    1  5  1             12625           -2122              6660
  92    d  o  1             -2784             355             -1408
  19   -7  3  1         -52656061        10293420         -28546726
       -2 -1  0          12439854        -1301678           6139812

1751    2  6  1                 9              -3                -8
  17    d  o  1                -1               1                 2
 103  -13  9  1             -2990            -867              -272
       -1 -1  0              -718            -209               -58

1752    1  3  1                17              16                 0
  24    d  o  1               -24              -4                -4
  73   -7  9  1                17              16                 0
        0  0  0               -24              -4                -4

1752   73  3  1             -2263           39712             38544
1752    d  o  4                16           -1884             -1864
  73   -7  9  1            683353         -905200          -3354496
        0  0 -2           -413000           99036             68236

1753
1753    n     1
1753  -10 48  1


1756  439  3  1      -13648347131       607928956       -1531177125
1756    n  o  5        651400068       -29014866          73079097
 439  -28 18  1        257465598       -32496097         -45393478
        0  0 -5         12257235        -1551471          -2169306

1757   14  5  4           -275835         -221382            -98392
1757    n  o  1              6587            5278              2352
   7   -1  3  1          -9494814         5755932          -6690656
        0  0 -1           -243628          123592           -165732
              c2             7896           -2618              5012
                             126            -112                98

1763    2  4  1           -298602          165948            -43640
  41    d  o  1            -67312           17656            -12536
  43    8  6  1           -298602          165948            -43640
        0  0  0            -67312           17656            -12536

1767    2  5  1             76952          -71612           -182038
  57    d  o  1            -33082           14812             15898
  31   -4  6  1 4730600890925134590 2 -29137558672655863500 -54080331396888750600
       -1 -2 -1 -8422987566315869160  4361369559608261720  6389054233920086800

1767   38  5  1            -86336          -10773             -8645
  57    d  o  1             11434            1427              1145
 589   41 15  3         -109707862        12233625          -3109350
        2  1 -1          -14681670         1624685           -390320
```

```
1767    38   3   1          -1444              95            152
  57    d    o   1            158             -17            -22
 589   -13  27   3          27702            3021           1634
         0   0  -1           2752             455            144

1767     2   5   4           1769           -4464           3162
  93    d    o   1          -1385            -232           -482
  19    -7   3   1           1769           -4464           3162
         0   0   0          -1385            -232           -482
                 c4           -78             -66              1
                             -14              -6             -3

1767    62   5   1         -77500           -9672          -7750
  93    d    o   1           8042            1002            804
 589    41  15   3         142538           37944          20460
         2   1   0         -32524           -1964          -2608

1767    62   3   1         -18259            1550           2480
  93    d    o   1          -1603             150            280
 589   -13  27   3         -18259            1550           2480
         0   0   0          -1603             150            280

1768     1   5   1           4716            2849           2073
 136    d    o   2           -807            -490           -355
  13     5   3   1    12645779597     -9973517228     3739637288
        -1  -2  -2     2180232074     -1703479784      646399326

1768    13   6   1         -40092           31031         -11921
1768    d    o   8           1907           -1476            567
  13     5   3   1       -5255367         2203812        8070920
        -1  -2  -4        -249974          105484         383962

1769     2   6   1             10              -6             -5
  29    d    o   1              2               0             -1
  61    -1   9   1       -3863059        -1406268        -678078
         0  -1  -1         717891          260964         125478

1769   122   6   1          16531          -39284           6771
1769    d    o   2           -393             934           -161
  61    -1   9   1       -8989936        -3267343       -1581486
         0  -1  -1         213444           77715          37224

1771     2   5   1           -134              51           -155
 253    d    o   1            -10               7             -3
   7    -1   3   1         -23021           14674         -10626
         1   2   0          -1307             606          -1238

1773     2   4   4           -225             788              0
 197    d    r   1            -65               4            -32
   9    -3   3   1           -526             360            -60
         0   0  -1            -24              20            -20

1777
1777    n       1
1777    14  48  16
```

1781	2	6	1	188	105	81
137	d	o	1	-16	-9	-7
13	5	3	1	-567178	-342500	-249477
	0	-1	-1	48336	29362	21253
1781	26	6	1	-13675181	10752313	-4047485
1781	d	o	2	324633	-254425	96167
13	5	3	1	-12827153314780530787871	692773601863496872	-37981747974137558266
	0	-1	-1	3039477346452629981	-2385979222854449472	899994703412023670
1784	1	4	3	-7389	-1494	-160
8	d	r	1	5458	1052	158
223	-28	6	1	-39	6	-10
	0	0	0	-58	-2	-8
1784	223	3	1	417233	284548	-126218
1784	n	o	1	-19746	-13476	5974
223	-28	6	1	1995181	1455298	750172
	0	0	-1	-338690	-16582	24528
1785	14	5	1	236131	-213612	183827
1785	d	o	8	-5589	5056	-4351
7	-1	3	1	-76449751	-27130215	22103655
	3	3	-4	1278843	1834507	1625387
1789	3578	4	1	1500390170624293	-9955740636378	98979722703058
1789	n	o	1	-35473091681113	235379374820	-2 340135817176
1789	-82	12	4	-5367	-3578	-1789
	-2	0	-1	1185	12	-35
1791	6	5	4	4782	-1791	-5373
597	n	o	1	-198	75	219
9	-3	3	1	-134319	53730	150444
	0	0	-1	5373	-2094	-6204
		c2		-462	210	513
				18	-6	-21
1791	1194	5	1	-2415220531421343	8109384008391	175121037542817
597	n	o	1	98848397631561	-331894998651	-7167227061669
1791	-84	6	3	90732465960	-848858778	-7594934301
	-2	-1	-1	-4300613676	-7833384	270236619
1791	1194	5	1	12235515	-718191	-1011915
597	n	o	1	-527151	27285	40761
1791	51	39	3	-933112299729297	23358791412630	-51512842457190
	1	-1	-1	38347194012003	-943421979210	2112187761330
1793	326	3	1	60671108407	-1051192542	12780901720
1793	n	o	1	-1432820493	24825164	-301836218
163	-25	3	4	79218	-25102	-30481
	2	4	-1	2598	-600	-571
1799	2	6	3	796	782	263
257	d	o	3	-50	-48	-15
7	-1	3	1	-44202	89436	164737
	0	1	-3	-2354	5906	10419
		b2		-2281	-1799	-771
				141	115	53

1801						
1801	n		1			
1801	74	24	1			
1807	2	6	1	2269	−399	127
13	d	o	1	−407	161	11
139	23	3	1	3679379	1286246	935402
	1	1	0	1535325	221738	271518
1807	26	6	1	1489725692	109801978	87413742
13	d	o	1	−413175922	−30453568	−24244214
1807	71	27	3	−1519445925322	−112420104160	−89291894796
	1	1	0	423162185880	31071032004	24792954540
1807	26	6	1	−958906	−73879	−48841
13	d	o	1	260972	19953	13097
1807	44	42	3	−14508	455	1235
	−1	−1	0	−4126	105	335
1812	1	3	1	74	−24	−40
12	d	o	1	−79	16	16
151	−19	9	1	−19007	4576	6048
	0	0	−1	10824	−2688	−3496
1817	158	5	1	−2215367296	773085784	−93832171
1817	n	o	1	51971862	−18136364	2201275
79	17	3	1	169834521372	−59274579191	7202471712
	−1	1	−1	−3984484012	1390533271	−168960098
1820	1	3	1	293	−224	42
140	d	o	2	−42	34	−20
13	5	3	1	293	−224	42
	0	0	0	−42	34	−20
1820	7	3	1	−6608	−1575	−630
140	d	o	2	729	291	6
91	−16	6	3	−6608	−1575	−630
	0	0	0	729	291	6
1820	7	5	1	−1036	−497	−483
140	d	o	2	−324	−64	−19
91	11	9	3	1057	−1820	−2870
	0	3	0	−1494	−82	220
1820	7	3	1	11333	8820	3150
1820	d	o	4	−552	−402	−162
7	−1	3	1	−1353863	808290	−777420
	0	0	−2	−60318	30804	−49218
1820	13	3	1	44759	27118	19656
1820	d	o	4	−2100	−1270	−922
13	5	3	1	−236951	191282	−69342
	0	0	−2	−11454	8744	−3430

```
1820    91  3  1        2279914       -786695       -985530
1820     d  o  4        -106863         36879         46208
  91   -16  6  3       30880031     -12319125     -14618695
         0  0 -2       -1688758        504773        670643

1820    91  5  1         533988        159341        109109
1820     d  o  4         -25034         -7470         -5115
  91    11  9  3      213206721     -53871090      25379900
         0  3 -2       10006504      -2525122       1187854

1821  1214  5  4       43441169       2255612       2798877
1821     n  o  1       -1017965        -52858        -65585
 607   -49  3  4      -64933825      -7385976       -615498
         2 -2 -1        1391273        173680          -598
           c3           -116544        -13354           607
                          2756           338            39

1827     2  6  1          -2624           782           724
  29     d  o  1           -324           140           190
  63    15  3  3      131129186     -47121984     -55008708
        -2 -1  0       25624632      -8812036      -9766952

1827     2  6  1            400           -38          -163
  29     d  o  1            -72             8            31
  63   -12  6  3          -1912           174           783
         1  2  0            348           -40          -149

1827     6  5  1          -3627          1284          4788
 609     d  o  2            147           -52          -194
   9    -3  3  1       -1554771      -1563912      -1116297
        -1 -2 -1         -54201        -69906        -41757

1827    42  5  1        -220752         12159        -78582
 609     d  o  2           8946          -493          3184
  63    15  3  3   -751151261721  263976866559  300348318501
        -2 -1 -1    -30439156755   10696944873   12170376213

1827    42  5  1       23911650      -7553616       2459982
 609     d  o  2        -969024        306100        -99646
  63   -12  6  3 -9710627741048816526 1908528964360343892 1838855249814483432
         1  2 -1 -47406006860572752 -31988717809577712 110111073098796072

1829    62  5  1          14136        -11253        -15345
1829     n  o  1           -330           263           359
  31    -4  6  1       -6008203       1415646      -2090547
        -2 -1 -1        -138141         31740        -51099

1832     1  4  1            101            22             8
   8     d  o  1             75            14             2
 229   -22 12  1          -2485         -1194           154
         0  0 -1          -5566          -634          -492

1832   229  4  1       -2539381       -499678       -122744
1832     d  o  2         115565         23928          6518
 229   -22 12  1      2492042807     503166586     130251078
         0  0 -1     -116497230     -23510024      -6098894
```

```
1833    2   5   4         1459         -564        -2538
141     d   o   1         -149           68          206
13      5   3   1      -875326       367164      1347396
        0   0  -1        73996       -31216      -113396
              c4          -800          320         1234
                           68          -30         -104


1833   26   5   4       -20800       -18330       -20163
1833    d   o   2          486          428          471
13      5   3   1        -9139         9165       -12831
        0   0  -1         -471          323           97
              c2           468         -299         -221
                            2           -3            9


1835    2   4   1           44           17            9
5       d   o   1           50            3            5
367    35   9   1        -7141         1134         -164
        0   0  -1         3423         -474          100


1841   14   5   1    -101308669     58026787    -64933421
1841    n   o   1      2361129     -1352389      1513357
7      -1   3   1    -12266569      7181741     -8580901
        2   1  -1       298741      -157065       204587


1843    2   6   1         -291          316          401
97      d   o   1           29          -32          -41
19     -7   3   1     -18650382     19870838     25552225
       -1   0  -1      1888888     -2020440     -2594993


1843  194   6   1   -81935803970    851043274   6284318824
97      d   o   1    -8345578676     84544814    636460580
1843   80  18   3
       -1  -1  -1


1843  194   6   1        14065         -388        -1067
97      d   o   1        -1439           38          107
1843   53  39   3     -56322177      3118259     -1359067
        1   0  -1      5717717      -315757       139179


1845    2   5   1          506          288          324
205     d   o   2           14           36           14
9      -3   3   1      -472318      -512500      -346040
       -1   1   0       -31160       -37152       -23444


1848    1   5   1         -814         -687         -278
264     d   o   2          100           85           35
7      -1   3   1   -6286133699  -5040951300  -2243287200
        3   3  -2    773759180    620523490    276177050


1848    7   5   1        -8008         4431        -5642
1848    d   o   4          374         -205          263
7      -1   3   1   1788765139   -998148228   1239213360
        3   3  -2    -83595252     46137914    -57786926


1852  463   5   1  -315414779312  33744300254  -17021223163
1852    n   o   1   14658567918  -1568230627    791043325
463    23  21   1  797862847533   5136797022 -363000095092
        1   2  -1  131339573084  -9845775298 -11783397214
```

1853	2	4	1	-38884824	-11045352	-5820200
17	d	o	1	-9444746	-2676476	-1414432
109	2	12	1	-2743317078994	638869542500	-434999143260
	0	0	-1	829741042936	-108308604800	130114818520

1853	218	4	1	-99299	17440	-16132
1853	d	o	2	2343	-414	358
109	2	12	1	-2199511	414200	995715
	0	0	-1	-61543	11450	21485

1855	2	6	1	-8351	4786	-5372
265	d	o	2	513	-294	330
7	-1	3	1	-286198	157410	-197955
	1	2	-2	17466	-9780	12093

1857	1238	3	1	97204996784	-5721550704	-14355349705
1857	n	o	1	-2255706746	132772398	333125459
619	17	27	1	3044861	-97183	-350973
	0	0	-1	-50225	4873	9725

1860	1	3	1	-31	15	-15
60	d	o	2	14	-1	5
31	-4	6	1	-2234	630	-825
	0	0	-2	655	-124	229

1861	3722	6	1	51157168575	-1496413351	2545375306
1861	n	o	1	1193253233	-34963877	59026258
1861	-37	45	1			
	-1	0	-1			

1865	2	6	1	4145	-330	444
5	d	o	1	1997	-124	208
373	-13	21	1	6769706737	-477460480	714773690
	-1	0	-1	3027589115	-213523464	319652294

1865	746	6	1	-52220	-1865	-373
1865	d	o	2	758	75	-39
373	-13	21	1	-9742128884	1150848605	1923646790
	-1	0	-1	226419272	-26708439	-44455660

1869	14	5	1	-6727	16121	29057
1869	d	o	2	163	-377	-667
7	-1	3	1	71197785071	-160082840400	-288562120602
	1	2	-1	-1649691585	3704443864	6672799094

1873	3746	6	1	241251985371735189	-14125605054477054	4658346083135374
1873	n	o	1	-5574449780993065	326390996869904	-107637316488448
1873	65	33	1	6426263	3710413	5032751
	-1	0	-1	-1753113	-54139	11105

1876	1	5	1	323	-18	-112
28	d	o	1	55	-31	-56
67	5	9	1	128924383189785	44371567010218	26175538658598
	-2	-1	-1	-48729253871046	-16770769628886	-9893497564608

```
1876    7   5   1                      -924                      -133                     -14
  28    d   o   1                      -365                       -46                       1
 469  -43   3   3               -3096305093                  18737782              -401162986
       -1  -2  -1               1168723782                  -7294384               151615574

1876    7   3   1                      1855                       -49                    -294
  28    d   o   1                      -700                        21                     112
 469   38  12   3                      3045                      3577                    5537
        0   0  -1                     15568                       945                    -203

1876    1   5   1                       -55                       -36                     -26
 268    d   o   1                        -5                        -5                      -1
   7   -1   3   1                     34171                     27604                   12194
        1   2   0                      4208                      3350                    1502

1876   67   5   1                    -17889                     -9380                   -4087
 268    d   o   1                      8180                       253                    -436
 469  -43   3   3                   3688417                   1335042                  500624
       -1  -2   0                  -1171148                    -55774                   51254

1876   67   3   1                    -17554                      1943                    -536
 268    d   o   1                      1845                      -271                      44
 469   38  12   3                    -17554                      1943                    -536
        0   0   0                      1845                      -271                      44

1881    6   5   1                  -3337938                  -3514098                -1787700
  33    d   o   1                   2645400                    179650                 -195136
 171   24   6   3  -11406722955009870187650  -2797802595596115327132  -410322823226803287108
       -1   1  -1   1985654047402551178752   4870352414084045592 48   71428414632295930896

1881    6   5   1                     -1863                      -543                    -315
  33    d   o   1                       369                        89                      43
 171   -3  15   3                -128195787                 -33931359               -18097101
       -1   1  -1                  22313049                   5907159                 3148875

1881    2   3   1                      -449                       187                     462
 209    d   o   1                        31                       -13                     -32
   9   -3   3   1                       563                       561                     396
        0   0  -1                        33                        43                      26

1881   38   5   1                   1919114                    -25878                  375212
 209    d   o   1                   -111568                      6998                  -25184
 171   24   6   3   35181654153705383989406  -1265360098326132666396  7364206346709443227260
       -1   1  -1  -2433724565131351610976    8755999670123084182 4  -509353888616107730672

1881   38   5   1                      4294                     -1558                     722
 209    d   o   1                      -380                       118                     -28
 171   -3  15   3             1774950654614              -219330870000            -469857074148
       -1   1  -1              122778071592               -15170902060             -32500389308

1883    2   6   1                      -943                      1595                    2374
 269    d   o   1                        21                       -77                    -170
   7   -1   3   1            -1541284368690              3463272857472             6240640860676
        1   1  -1              93976539640               -211161031540             -380496664812
```

```
1884    1   3   1              74           -12              -3
  12    d   o   1             -15             4              -7
 157   14  12   1              74           -12              -3
   0    0   0                 -15             4              -7

1884  157   3   1         2037860       -203472         -634437
1884    d   o   2          -93889          9378           29235
 157   14  12   1        10228864       2437425         1627776
   0    0  -1             477627        111147           75468

1885    2   4   1            1854          -783           -2842
 145    d   o   4            -154            65             236
  13    5   3   1           -1912         -1276           -1073
   0    0  -4               -178           -98             -59

1885   26   3   1         -173017       -104806          -76154
1885    d   o   4            3985          2414            1754
  13    5   3   1           16614        -31668            3016
   0    0  -2               1000          -356             340

1891    2   4   1             -29             5              -9
  61    d   o   1               3            -1               1
  31   -4   6   1             175           -32              67
   0    0  -1                 -21             6              -7

1891  122   6   1           44591          -122           -3538
  61    d   o   1            5917          -122            -488
1891   83  15   3   42963857207960899   -355046161727458   -3207904179737938
  -1   -1  -1      5500964637637573    -45459423702174    -410729989247494

1891  122   6   1           -7564          -549             -61
  61    d   o   1             838            65               9
1891  -79  21   3       -674057503      44216460        52262482
  -1    0  -1          -86353769       5656508         6691742

1892    1   3   4            -243            66             -44
  44    d   r   1             -52            28              -8
  43    8   6   1               8            -1               2
   0    0   0                   1            -1               0

1893 1262   5   1      5007184396    -121175978       512307007
1893    n   o   1      -115084700       2785082       -11774865
 631  -43  15   1   -116877445396618   12856309492428   15212385147468
   1    2  -1      -2448256153700    289727912888    373996639252

1895    2   4   1            5367          -790             218
   5    d   o   1            2661          -314             126
 379   29  15   1            5367          -790             218
   0    0   0               2661          -314             126

1896    1   5   1              47            -9              -7
  24    d   o   1               2            -3               4
  79   17   3   1       810811009      -36376812      -332350044
   1    2  -1        -332160794       14517578       135384208
```

1896	79	5	1	10413701	3021513	2694137
1896	d	o	2	-478322	-138781	-123746
79	17	3	1	-1340083247873	425368768884	-71150880108
	1	2	-1	-55469390946	21302697146	-1694418056
1897	14	5	1	12123197324	-26812529919	-47942076336
1897	n	o	5	-278345054	615607821	1100736008
7	-1	3	1	-12897689	-10469543	-4786131
	1	2	-5	-298027	-236097	-102163
1897	542	5	1	-54716988326	5318632721	14822364783
1897	n	o	5	1256286000	-122114247	-340317147
271	29	9	1	-2837648455516501	481877373970755	-102112386014709
	1	2	-5	-65155762315173	11063919397419	-2343576744531
1897	3794	5	1	2771665639560202	-157403739951738	-224619254468462
1897	n	o	5	-63636629966094	3613943692060	5157192186478
1897	-55	39	3	-1145442746	-64209656	-32620812
	-2	-1	-5	-16141628	-1707324	-203120
1897	3794	5	1	-30986170871080446	1049793060820238	2546774783691250
1897	n	o	5	711433392846788	-24102940700666	-58473201891142
1897	26	48	3	-4426502789552317868338	206842597705754237360	-145055258573613775604
	-2	-1	-5	101632119806561708552	-4749076340046635608	3330351030961554272
1899	6	5	1	145974	-68232	65064
633	n	o	1	-5802	2712	-2586
9	-3	3	1	-7078408554	1028126196	5881286556
	-1	1	-1	-189641736	145141452	301822200
1899	1266	3	1	7344699	501969	8862
633	n	o	1	-291813	-19959	-360
1899	87	3	12	6011601	-415881	-427275
	-2	2	-1	244971	-16395	-16707
1899	1266	3	1	-708183395258910	18198504416256	55234889786928
633	n	o	1	28147752503352	-723326425896	-2195390318520
1899	-48	42	12	-166221823494	-4611425256	1802872620
	4	2	-1	-203086656	-347845152	-427795512
1905	254	3	1	331867119094	81120403164	67313423412
1905	d	o	2	-7603555784	-1858585776	-1542247904
127	20	6	1	-3176996116150	804054325848	-167106494844
	0	0	-1	-72394142704	18518706416	-3748444256
1908	3	3	1	-1536	1152	-594
636	d	o	2	123	-90	48
9	-3	3	1	-23109	17010	-9342
	0	0	-1	1836	-1374	708
1919	2	6	1	260	-121	-35
101	d	o	1	8	7	21
19	-7	3	1	-12219	33128	57974
	-1	-1	0	5559	-3936	-3526

```
1928    1  6  3              -10953                       1649                         -813
   8    d  o  1               -7757                       1164                         -576
 241   17 15  1            -2565543                     385604                      -190932
       -1  0  0            -1815658                     272910                      -134468
             b2                1429                       -140                         -232
                             -499                          22                          202

1928  241  5  1               40729                      -6989                         2169
1928    d  o  2               -2125                        266                         -134
 241   17 15  1       -199169034971               -39106169624                 -26061621428
       -1  0 -1         -9131820220                -1772230982                  -1191523062

1929 1286  3  3   23870310520281078         -697537633590656          2343570634385280
1929    n  o  3    -543490497009256           15881866088032           -53359522398176
 643  -40 18  1          1246739706                 -128229632                 -153579264
        0  0 -3            24099968                   -2794336                   -3917600
             b1          1699524874                 -185019392                 -240080768
                         38695592                    -4212608                   -5466272

1932    7  5  1               -3815                      -3689                        -2590
1932    d  o  4                 198                        113                           19
   7   -1  3  1       8056772057677               6461691386388                2876385028398
        0  3 -2       -366615577212               -293972750930                -130800068158

1933
1933    n     1
1933   62 36  1


1935    2  6  1                4154                        628                          607
   5    d  o  1                1796                        290                          271
 387  -39  3  3       -117788617093                17487460410                  -915396870
        2  1 -2         52676926185                 -7820575282                   409443478

1935    2  4  4                 748                        -35                           75
   5    d  r  1                 268                        -25                           31
 387   15 21  3                -131                          6                          -36
        0  0 -2                -129                         10                           -4

1935    6  5  1               -7083                      -8769                        -6120
 645    d  o  2                 315                        331                          200
   9   -3  3  1       52671543344601               59895936296010               39094266931200
       -2 -1 -1       -2073942830445               -2358398124906               -1539336193176

1935  258  5  1              -72369                     -11223                       -11352
 645    d  o  2                2709                        463                          446
 387  -39  3  3       -135931259397                2747386530                  19266218370
        2  1 -1         4405827945                  32336934                   -765962610

1935  258  3  4             5868468                    1004265                       399255
 645    d  r  2             -224460                     -39989                       -15013
 387   15 21  3             -682539                      48762                       -74304
        0  0 -1              -27993                       1782                        -2952

1937    2  6  1               20380                     -14782                            3
 149    d  o  1               -1130                        984                         -829
  13    5  3  1   71962926109082633        -30272372339828676       -110526791644141442
        0 -1 -1    5895475852474651         -2479983186689012         -9054688564542510
```

1937	26	6	1	−25610	18967	−7709
1937	d	o	6	582	−431	175
13	5	3	1	−27092	13559	30992
	0	−1	−3	−352	99	782
1939	2	6	1	979	−1877	−3536
277	d	o	1	47	−125	−216
7	−1	3	1	8944609	−20233188	−36411650
	1	1	0	542763	−1211476	−2185846
1939	554	6	1	−330738	−27700	−4432
277	d	o	1	24526	1586	526
1939	−67	33	3	327069966	−5912288	18493628
	1	1	0	19714288	−357940	1106988
1939	554	6	1	−39832334911	−2966813209	−1901783388
277	d	o	1	2393294135	178258477	114267120
1939	41	45	3	454988145237	−26788577482	12147867132
	1	1	0	31171648689	−1323999210	912950676
1943	2	6	1	−530	112	194
29	d	o	1	−54	12	44
67	5	9	1	−3790762	820352	1995084
	−1	0	0	−706880	151268	369876
1948	487	5	1	53029887293	−2270854702	5324898421
1948	n	o	1	−2403013571	102902249	−241294180
487	−25	21	1	−13126619915	1315025646	2060043116
	−1	−2	−1	−532006012	68638114	96279358
1953	6	5	1	70815	13089	768
21	d	o	1	−15459	−2855	−166
279	33	3	3	−89549137623	−17098689258	−504279972
	1	−1	−1	20777190357	3717901782	325059252
1953	6	3	1	−35940	2505	6165
21	d	o	1	−6024	295	1457
279	−21	15	3	132405873	−7728624	−26355510
	0	0	−1	28382205	−1789800	−5823510
1953	42	5	1	−771792	36771	28602
21	d	o	1	82530	−6931	−11450
1953	78	24	9	1804614	330813	150507
	−2	−1	−1	1040298	33003	−14577
1953	42	5	1	6216	−357	189
21	d	o	1	−1806	95	−29
1953	−57	39	9	−127392570921	−9715557726	−6860747628
	1	2	−1	27803928117	2120006790	1496784228
1953	42	5	1	2936155467	−137268306	91353759
21	d	o	1	−640753407	29955898	−19936105
1953	−30	48	9	12692867334	−395868501	−989410842
	1	2	−1	2773618050	−86089689	−215729220

```
1953   42   5   1         -512043              19614              44688
  21   d    o   1          128751              -4960              -9098
1953   -3  51   9  -229742093555121     8826699405888     18006901694838
       -1  -2  -1    50131805135811    -1926302839464     -3929509206786


1953    6   5   1           24510               9825               1191
  93   d    o   1           -2550              -1017               -123
  63   15   3   3        -10051527           -4021506            -489366
       -2  -1   0         1042809             416694              50202


1953    6   3   1              60                -18                -45
  93   d    o   1             -12                  0                  3
  63  -12   6   3              60                -18                -45
        0   0   0             -12                  0                  3


1953  186   5   1        -31343790             394134           -1914126
  93   d    o   1          3272856             -42258             196782
1953   78  24   9           622914             -10044              39060
       -2  -1   0           -69936                672              -4104


1953  186   5   1           138570              -7626               3162
  93   d    o   1           -14694                780               -342
1953  -57  39   9        138660954           10625436            7533000
        1   2   0         14486424            1099332             773076


1953  186   5   1       3843408210         -179683347          119581725
  93   d    o   1        398774886          -18643143           12407271
1953  -30  48   9          -103602               5301              -3069
        1   2   0           -11346                483               -369


1953  186   5   1         -6460617             258075            -247752
  93   d    o   1          -668949              26739             -25752
1953   -3  51   9     -17996449809          709671654         -691590222
       -1  -2   0      -1849595439           74886858          -71053098


1953    2   5   1             194                 10                107
 217   d    o   1               2                -12                 -1
   9   -3   3   1         -100903            -131068             -77903
       -2  -1   0            8029               8022               5755


1953   14   5   1            -749               -294                -42
 217   d    o   1              49                 20                  2
  63   15   3   3           -4543              -1519               -434
       -2  -1   0             217                107                 -4


1953   14   3   1        -16258214            4963700           -1250900
 217   d    o   1           994000            -325680             130840
  63  -12   6   3        -16258214            4963700           -1250900
        0   0   0           994000            -325680             130840


1953   62   5   1            54715              10044                558
 217   d    o   1            -3689               -686                -42
 279   33   3   3          -988807            -202027             -21266
        1  -1   0            74431              12439                 84
```

```
1953    62  3   1            -2945               -496                 -341
 217     d  o   1             -155                -38                  -27
 279   -21 15   3            -2945               -496                 -341
         0  0   0             -155                -38                  -27

1953   434  5   1        -4504156594          276754422            340851882
 217     d  o   1         -305762548           18787318             23138506
1953    78 24   9  -14027672476046950    -1028862091110812    -178622423158060
        -2 -1   0     952180191924976       69848574890984      12131567708384

1953   434  5   1            -705901              37975               -15841
 217     d  o   1              47957              -2585                 1075
1953   -57 39   9            4379711             346115               239134
         1  2   0             310093              22823                16472

1953   434  5   1         104364103690        -4879340914          3248592424
 217     d  o   1          -7086550772          331289398          -220387800
1953   -30 48   9       -1493805853414        46478352452         116311194496
         1  2   0        -101402016800         3154908640          7895840112

1953   434  5   1            -565502              22568               -19313
 217     d  o   1             -35588               1424               -1531
1953    -3 51   9          -54669895            2178463            -2109457
        -1 -2   0           -3714389             148835             -142111

1956     1  3   1              -1446                380                 406
  12     d  o   1               -808                219                 240
 163   -25  3   4             -30863              -6354                -174
        -2  2  -1              16498               4022                 484

1957    38  3   1           24564302          -26223040           -33700148
1957     n  o   3            -555276             592772              761792
  19    -7  3   1              29393             -26638              -35834
         0  0  -3               -531                674                 830

1957   206  5   1       -13307246401         -3866067405         -1143085966
1957     n  o   3         300810229            87392481             25839434
 103   -13  9   1      -35953738998510     -10445423067732     -3088435260256
        -1  1  -3        812735662776        236118298044         69813022548

1957  3914  3   1        5544262013929       -35183096689       -402259275580
1957     n  o   3         -125327863227        795312761          9093057894
1957    86 12  12               5871               3914                1957
         2  4  -3               1325                 12                  49

1957  3914  5   3         -70701930427        2783983189          5757110428
1957     n  o   3          1598217417          -62931951          -130139504
1957     5 51   3          243750221           -9871108             9608870
        -2 -1  -3            5652753            -228492              205862
                b1         -2005925              84151              -66538
                            -41761               1761               -1796

1961     2  6   1              -6874              -2634               -2156
  53     d  o   1               -892               -372                -322
  37    11  3   1          540424678          216019096          182027652
         0 -1   0           74260040           29669564           24985652
```

```
1961   74  5  1            -13187133              1383097               8780026
1961    d  o  2               297791               -31233               -198270
  37   11  3  1           1200618402          -1695378628           -3916852375
        0 -1 -1           -167598714            -17857446             41156423

1963    2  6  1              4550196             -1112875             -1459403
  13    d  o  1             -1262774               308699               404615
 151  -19  9  1 -16897775613265798639 -4951534239937865658 -1632510049095120364
        1  0 -1 -5653674480545386137 -1136840188341046926 -142756065239858172

1963   26  6  1              4729647               319553                14144
  13    d  o  1              1298533                87733                 3884
1963  -88  6  3             19266572                11427              1391221
       -1  0 -1             -5923246                36967              -343857

1963   26  6  1                -9386                 -741                 -325
  13    d  o  1                -2820                 -197                  -99
1963   -7 51  3          29701571497           -836150458            7441382936
       -1  0 -1          -36512476951           1435065558            237360256

1967    2  4  1                  138                  101                   30
 281    d  o  1                   -8                   -7                   -4
   7   -1  3  1                  138                  101                   30
        0  0  0                   -8                   -7                   -4

1976    1  6  1                -6241                -1133                  -65
   8    d  o  1                -4354                 -811                  -54
 247  -31  3  3          -8495970185839         89401194364          -1475167934756
       -1  1 -2           5945437211018         -74639417970          1042392969272

1976    1  4  1                   89                   -2                   10
   8    d  o  1                  -30                    8                   -4
 247   -4 18  3                 8279                 -852                  986
        0  0 -2                -5960                  596                 -692

1976    1  6  1                  -89                  125                  170
 104    d  o  2                   28                  -24                  -29
  19   -7  3  1              -397227               426244               546832
       -2 -1  0                78402               -83268              -107198

1976   13  6  1                 8931                 1599                  143
 104    d  o  2                -1656                 -315                  -12
 247  -31  3  3             -4211519               419796             -1044576
       -1  1  0             -1571034               -54684              -213354

1976   13  4  1                  195                  -26                  -26
 104    d  o  2                  -14                    2                    8
 247   -4 18  3                  195                  -26                  -26
        0  0  0                  -14                    2                    8

1976    1  5  1                 -152                 -125                 -131
 152    d  o  1                  -35                  -16                   -5
  13    5  3  1              -202691              -176776              -193116
       -1 -2  0               -49956               -21494                -5102
```

```
1976   19  5  1          170449           -46227           -44517
 152    d  o  1           36026            -5959            -7126
 247  -31  3  3       -970556537        209648660        221394384
       -1  1  0       -161156826         33326936         35872610

1976   19  3  1             -551             -228              -38
 152    d  o  1              218               24               22
 247   -4 18  3             -551             -228              -38
        0  0  0              218               24               22

1976   13  5  1         -6370624          5000905         -1886365
1976    d  o  2           286627          -225002            84871
  13    5  3  1    45169645613521   -36950735504644   20880649866024
       -1 -2 -1    -2400951532422     1817579119052    -373227976678

1976   19  5  1           -30837           -18373            -3686
1976    d  o  2             1388              826              165
  19   -7  3  1      -56047101949      -32417828196      -7193208100
       -2 -1 -1       2521651626       1458536134        323593488

1976  247  5  1          -277875             3705           -48165
1976    d  o  2            12612             -147             2168
 247  -31  3  3    -18880047552153      216987252300    -3295647072900
       -1  1 -1      849805326390      -9835181370     148200921540

1976  247  3  1            68419            11856             6422
1976    d  o  2            -3074             -534             -290
 247   -4 18  3            88673            15314             5928
        0  0 -1            -3358             -772             -410

1980    1  5  1            -1985              766             2327
 220    d  o  2             -280              113              309
   9   -3  3  1          3723721         -1469600         -4240390
       -3 -3  0           503140          -198842          -571350

1981   14  5  1         -6288373          3150707         -4268180
1981    n  o  1           141285           -70789            95896
   7   -1  3  1       -107475179        248492678        445269370
        2  1 -1         -2517843          5500310          9967362

1981  566  5  1       5883872810        987928379        875907357
1981    n  o  1       -132196828        -22196435        -19679585
 283   32  6  1        -25120495          1196524          4170005
       -2 -1 -1           382429             4982           -97801

1981
1981    n     1
1981  -43 45  3


1981 3962  5  1   47699081768828    3567254836495    1911277639222
1981    n  o  1    -1071686551850     -80147853869      -42941928164
1981   11 51  3  -928289679937123   38737334173972  -33573220738890
        2  1 -1    -20852094209083     870666095836     -754135275722
```

```
1985    2   4   1           -655            -103            -412
   5    d   o   1           1241             -93              12
 397  -34  12   4          -2362             395             465
       -2  -2  -1          -1204             155             203

1985  794   4   1   -11533792605040    1692306284104    2096962514856
1985    d   o   2     258876150210      -37983829232      -47066326104
 397  -34  12   4    -85408170107574   12532333293520   15528606002260
       -2  -2  -1     1917098113320     -281271256368    -348536345752

1988    1   3   3            -97              50             -58
 284    d   o   1             10              -6               8
   7   -1   3   1            -97              50             -58
        0   0   0             10              -6               8
            b2               76             -34              50
                             -9               4              -6

1989    2   6   1           -625              16             182
  17    d   o   1            153              -4             -44
 117  -21   3   3         -10453             306            3009
       -1  -2   0           2499             -56            -727

1989    2   6   1         915758          128870          -24104
  17    d   o   1         -65172          -60346          -44192
 117    6  12   3   -23786967794422   4409402417940    7584540913296
       -1   1   0   -5769204304224    1069431637672    1839519080384

1989    2   3   1            284             247              65
 221    d   o   2             -8             -21             -17
   9   -3   3   1          49883           66924           49530
        0   0  -2          -4251           -4148           -2314

1989   26   5   1          67574           19019           16861
 221    d   o   2          -4316           -1285           -1201
 117  -21   3   3   -17419814365993   -4758453881688   -4468913174622
       -1  -2  -2    1103057264361     338198371712     298833957862

1989   26   5   1           2899            1014             468
 221    d   o   2           -169             -60             -28
 117    6  12   3       21840572         -4067505         -6982053
       -1   1  -2       -1474512          272057           468931

1993 3986   6   1   29544291269827   -1045690080025    1238827929304
1993    n   o   1    -661789583177     23423367847      -27749639058
1993  -13  51   1   -1076312321739    -80398028565      -36800013569
        1   1  -1     24112423125      1800801659         824494593

1995    2   3   1            452             255              60
 105    d   o   2            -44             -25              -6
  19   -7   3   1           1352             750             165
        0   0  -2           -126             -76             -17

1995   14   3   1          13846           -1155           -4550
 105    d   o   2          -1352             113             444
 133   17   9   3         -23926            4655           11970
        0   0  -2           3896             -67            -880
```

```
1995   14   3   1          170973698         -18299568          31210116
 105    d   o   2          -16687280           1785344          -3045984
 133  -10  12   3       -646569230734       69235014324       -117812707152
   0    0  -2         63021984000        -6739515520        11524514944

1995    2   5   1             -16940            -13499             -5926
 285    d   o   2               1000               807               364
   7   -1   3   1    65882794187927     5289695834490     23497722589830
   0    3  -2      -3908065611679     -3130289962910     -1395695017342

1995   38   3   7              60097             20235             12160
 285    d  r2   2              -4929              -889              -826
 133   17   9   3              -3667               380              1140
   0    0  -2                195               -12               -68

1995   38   3   1               -266               171               228
 285    d   o   2                 24               -11               -12
 133  -10  12   3              -3838              1083              2451
   0    0  -2                468               -91              -101

1996  499   3   1        189209356932        -21307485129        7559307587
1996    n   o   5          -8470174095         953854035         -338401083
 499   32  18   1            44424972          -4977026           1797897
   0    0  -5            -1974801            224226            -79791

2007    6   5   1             192750           -143055             76098
 669    n   o   1              -7452              5531             -2942
   9   -3   3   1       -174246060135       129317404302       -68758808472
  -2   -1  -1          6733899111        -4998576438         2661602496

2007 1338   5   1          182155989          -12978600         -15393021
 669    n   o   1           -7042563            501782            595129
2007   69  33   3        -916839001065       23527298340       -26983709586
   1    2  -1         -20807460777          84595548         -2127005370

2007 1338   5   1         1932567060          149039820          67197705
 669    n   o   1          -74717934          -5762188          -2597969
2007   15  51   3    -8335529431964337     290464450096050    -353042676077400
   1    2  -1       -322256984984625     11231023734210     -13648958279280

2013    2   3   4                310               -33                66
  33    d   r   1                -34                13                -8
  61   -1   9   1                 97               -29                21
   0    0  -1                -21                 3                -5

2013  122   3   4             838201           -225456           -452925
2013    d   r   2             -17129              5590             10367
  61   -1   9   1             207217             76738             35502
   0    0  -1               4729              1690               846

2015    2   6   1                347                44                 2
   5    d   o   1                127                24                 6
 403  -37   9   3              35687              6370               130
   1   -1   0              21157              2734               742
```

2015	2	6	1	61487	−4250	−11391
5	d	o	1	27509	−1898	−5093
403	17	21	3	−31982343	2195710	6004740
	1	2	0	−14595187	1013710	2666524
2015	2	6	1	−8342308	−4008498	−1580484
65	d	o	2	1035986	496494	194884
31	−4	6	1			
	−1	1	−2			
2015	26	6	1	6214	988	169
65	d	o	2	−748	−126	−25
403	−37	9	3	−15570464	267930	−2002585
	1	−1	−2	−1823274	49488	−246055
2015	26	6	1	100386	−6890	−18538
65	d	o	2	−12404	862	2304
403	17	21	3	665422446	104411580	65749840
	1	2	−2	83011532	12917776	8067180
2017						
2017	n		1			
2017	−34	48	1			
2021	86	3	1	−81829	31347	−14018
2021	n	o	3	1821	−697	312
43	8	6	1	−1032	989	−129
	0	0	−3	58	−9	11

References

[1] Bauer, H.: Numerische Bestimmung von Klassenzahlen reeller
     zyklischer Zahlkörper - J. Number Theory 1 (1969), 161 - 162

[2] Billevič, K.K.: On units of algebraic fields of third and fourth
     degree (Russian) - Mat. Sbornik, n. Ser. 40 (1956), 123 - 136

[3] Cassels, J.W.S.: On a conjecture of R.M. Robinson about sums of
     roots of unity - J. reine angew. Math. 238 (1969), 112 - 131

[4] Châtelet, A.: Arithmétique des corps abéliens du troisième degré
     - Ann. sci. École. norm. sup. (3) 63 (1946), 109 - 160

[5] Cohn, H. and Gorn, S.: A computation of cyclic cubic units - J. Res.
     nat. Bur. Standards 59 (1957), 155 - 168

[6] Gras, G. et Gras M.N.: Nombre de classes des corps quadratiques
     réels $Q(\sqrt{m})$, m < 10 000 - Université de Grenoble 1971/1972

[7] Gras, G. et Gras, M.N.: Calcul du nombre de classes et des unités
     des extensions abéliennes réelles de Q - Bull. Sc. math. II Sér.
     101 (1977), 97 - 129

[8] Gras, M.N.: Nombre de classes, unités et bases d'entiers des
     extensions cubiques cycliques de Q - Bull. Soc. math. France 37
     (1974), 101 - 106

[9] Gras, M.N.: Méthodes et algorithmes pour le calcul numérique du
     nombre de classes et des unités des extensions cubiques cycliques
     de Q - J. reine angew. Math. 277 (1975), 89 - 116

[10] Gras, M.N.: Table numérique du nombre de classes et des unités des
     extensions cycliques réelles de degré 4 de Q - Publ. math. Fac. Sci.
     Besançon 1977/1978, fasc. 2

[11] Gras, M.N.: Classes et unités des extensions cycliques réelles de
     degré 4 de Q - Ann. Inst. Fourier 29, 1 (1979), 107 - 124

[12] Gras, M.N., Moser, N. et Payan, J.J.: Approximation algorithmique
     du groupe des classes de certains corps cubiques cycliques - Acta
     arithmetica 23 (1973), 295 - 300

[13] Hasse, H.: Arithmetische Bestimmung von Grundeinheit und Klassen-
     zahl in zyklischen kubischen und biquadratischen Zahlkörpern -
     Abh. Deutsch. Akad. Wiss. Berlin, math. -naturw. Kl. 1948, No 2
     (1950)

[14] Hasse, H.: Über die Klassenzahl abelscher Zahlkörper - Akademie
     Verlag, Berlin, 1952

[15] Hasse, H.: Vorlesungen über Zahlentheorie - Springer Verlag, 1964

[16] Ince, E.L.: Cycles of reduced ideals in quadratic fields - Brit. Assoc. Advancement Sci., Math. Tables IV, 1934

[17] Latimer, C.G.: On the units in a cyclic field - Amer. J. Math. 56 (1934), 69 - 74

[18] Leopoldt, H.-W.: Über Einheitengruppe und Klassenzahl reeller abelscher Zahlkörper - Abh. Deutsch. Akad. Wiss. Berlin, math.-naturw. Kl. 1953, No 2 (1954)

[19] Leopoldt, H.-W.: Über ein Fundamentalproblem der Theorie der Einheiten algebraischer Zahlkörper - Bayer. Akad. Wiss., math.-naturw. Kl. S. -Ber. 1956, No 5 (1957), 41 - 48

[20] Leopoldt, H.-W.: Über die Hauptordnung der ganzen Elemente eines abelschen Zahlkörpers - J. reine angew. Math. 201 (1959), 119 - 149

[21] Loxton, J.H.: On a cyclotomic diophantine equation - J. reine angew. Math. 270 (1974), 164 - 168

[22] Masley, J.M.: Class numbers of real cyclic number fields with small conductor - Compositio math. 37 (1978), 297 - 319

[23] Pohst, M.: Berechnung kleiner Diskriminanten total reeller algebraischer Zahlkörper - J. reine angew. Math. 278/279 (1975), 278 - 300

[24] Shanks, D.: A survey of quadratic, cubic and quartic algebraic number fields (from a computational point of view) - Proc. 7th Southeast. Conf. Comb., Graph Theory, Comput., Baton Rouge 1976 (1976), 15 - 40

[25] Stender, H.-J.: Über die Einheitengruppe der reinen algebraischen Zahlkörper sechsten Grades - J. reine angew. Math. 268/269 (1974), 78 - 93

[26] Weber, H.: Lehrbuch der Algebra, II Bd - Braunschweig, 1899

[27] Yokoi, H.: On unit groups of absolute abelian number fields of degree pq - Nagoya math. J. 16 (1960), 73 - 81

[28] Zimmer, H.G.: Computational Problems, Methods and Results in Algebraic Number Theory - Lecture Notes in Mathematics, Vol. 262, Springer Verlag (1972)

Terminology and notation

$\square$	: the end of a proof
#M	: the number of elements in a set M
<M>	: the subgroup generated by a subset M of a given group
$\sqrt{x}$	: for a real number $x \neq 0$, $\sqrt{x}$ lies on the positive part of the real or imaginary axis
$p^{\nu} \parallel c$	: indicates that $p^{\nu}|c$, $p^{\nu+1}\!\!\not|\,c$ for p prime and $c \in \mathbb{Z}$
$\overline{\Phi}_k$	: the multiplicative group of prime residue classes mod k
$\varphi(k)$	: $\#\overline{\Phi}_k$
ζ_k	$= e^{2\pi i/k}$
ρ	$= (-1 + \sqrt{-3})/2$
K_n	: a real cyclic extension of degree n over the rationals $\mathbb{Q}$ $(n|6)$; $K_1 = \mathbb{Q}$
f_n	: the conductor of K_n ; $f_1 = 1$
f_*, f_n'	: $f_* = \gcd(f_2, f_3)$; $f_n' = f_n/f_*$ $(n = 2,3)$
m	: $f_2 = m$ if $m \equiv 1 \bmod 4$; $f_2 = 4m$ if $m \equiv 2,3 \bmod 4$
a, b, ϕ	: $f_3 = (a^2 + 3b^2)/4$, $\phi = (a + b\sqrt{-3})/2$; the choice of a and b: p. 6 ; the decomposition of ϕ: p. 37
d_n	: the discriminant of K_n, pp. 6, 13
h_n	: the class number of K_n (in the ordinary "wide" sense)
h_R	: the relative class number of K_6, p. 58
$S_{m/n}$	: the trace from K_m to K_n
$N_{m/n}$	: the norm from K_m to K_n
G	: the Galois group of K_6, cyclic and of order 6
σ	: a generator of G, p. 12
$\gamma^{(i)}$	$= \sigma^i(\gamma)$ $(i \in \mathbb{Z}$, $\gamma \in K_6)$; also γ, γ', γ'', γ''', γ^{iv}, γ^{v}
s	: the integer defined on p. 12 ; cf. also p. 40
S	: the automorphism $\zeta_{2f_6} \mapsto \zeta_{2f_6}^{s}$ of the field $\mathbb{Q}(\zeta_{2f_6})$
$\mathcal{O}_n$	: the ring of integers of K_n
U_n	: the multiplicative group of units of K_n
Sr	: the signature rank of U_6, p. 60

U_R : the group of relative units of K_6
$$= \{\varepsilon \in U_6 \mid N_{6/3}(\varepsilon) = \pm 1, \ N_{6/2}(\varepsilon) = \pm 1\}$$

μ : the fundamental unit of K_2

τ, τ' : fundamental units of K_3, $N_{3/1}(\tau) = 1$, the choice of τ: p. 64

ξ : the number λ in [14, p. 21], defined here on pp. 16, 32

η : the cyclotomic unit in K_6, p. 16

ξ_A : $\xi_A = \xi$ if $\xi \in K_6$, $\xi_A = \eta$ if $\xi \notin K_6$; cf. p. 51

u, v, w : $N_{6/3}(\xi_A) = \pm\tau^u\tau'^v$, $N_{6/2}(\xi_A) = \pm\mu^w$, pp. 16, 17

ξ_R : a generating relative unit of K_6, p. 28

U_6^* : $= \langle -1, \mu, \tau, \tau', \xi_A, \xi_A', \xi_R, \xi_R' \rangle$, p. 17

ξ_0 : a candidate for ξ_R defined on p. 29

ξ_1 : $\xi_1^{2^n} = \pm\xi_0$ for some $n \geq 0$, $\pm\xi_1$ is not a square in K_6 ; for $n = 0$, $\xi_1 = \xi_0$; p. 29

ξ_B : in the case $N_{6/2}(U_6^*) \neq U_2$ the solution of $x^3 = \mu\xi_R\xi_R'$ or $x^3 = \mu^{-1}\xi_R\xi_R'$ in U_6 if it exists, p. 18

ξ_C : in the case $\langle -1 \rangle N_{6/3}(U_6^*) \neq U_3$ the positive solution of $x^2 = |\tau\xi_R|$ or $x^2 = |\tau'\xi_R|$ or $x^2 = |\tau\tau'\xi_R|$ in U_6 if it exists, p. 19

$\mathcal{M}$: the mean square modulus function, p. 23

α : the set defined on pp. 16, 32 ; $1 = \#\alpha$

A_t : the coefficients in Bergström's product formula, p. 33

$H, H^{(d)}, \bar{H}, \bar{H}^{(d)}$: the multiplicative groups of residue classes defined on pp. 32, 33, 35 ; cf. p. 40

$m(d), \bar{m}(d)$: indices of $H^{(d)}, \bar{H}^{(d)}$ in the corresponding full multiplicative groups of residue classes, pp. 35, 40

$\bar{H}_{-1-f}, \bar{H}_{-1}$: certain factor groups of $\bar{H}$, p. 35

character : a character χ always means a Dirichlet character modulo a natural number and it is assumed to be primitive, i.e. $\chi(x) = 0$ if and only if $(x, f(\chi)) > 1$ where $f(\chi)$ is the conductor of χ

$\tau(\chi)$: the Gaussian sum for the character χ

$\pi(\chi', \chi'')$: the Jacobi sum for the characters χ', χ'', pp. 44, 53

χ_n : a generating character of K_n ; the choice of χ_3: p. 9

π_i', $3\rho^\alpha$: the factors in the decomposition of ϕ, p. 37

$\left(\frac{\pi'_i}{\pi'_i}\right)_3$, $\left(\frac{\rho}{\cdot}\right)_3$: the cubic characters defined on p. 38

χ'_3 : a product of these characters which coincides with χ_3 or $\overline{\chi}_3$, p. 37 ; in Section 2 χ'_3 is denoted by χ

π_i : $\pi_i = \pi'_i$ if $\chi'_3 = \chi_3$; $\pi_i = \overline{\pi}'_i$ if $\chi'_3 = \overline{\chi}_3$; p. 42

$\chi_{n,k}$: for $n = 2$ or 3, $k|f_n$, $(k,f_n/k) = 1$, $\chi_{n,k}$ is that character with conductor k which appears in the decomposition of χ_n, p. 38

$\chi'_{2,4}$, $\chi'_{2,8}$: $\chi'_{2,4}(x) = (-1)^{(x-1)/2}$, $\chi'_{2,8}(x) = (-1)^{(x^2-1)/8}$

ψ : the character of $\overline{H}$ defined on p. 33 ; if possible, ψ is considered as a character of H, see pp. 34 - 36

θ, θ', θ'' : the Gaussian periods, multiplied by ±1, for the cubic character χ_3, pp. 7, 9

decomposable : p. 37, cf. p. 51

co-ordinates : the co-ordinates of a number $\gamma \in K_n$ are the rational numbers x_i in the representation $\gamma = x_1\omega_1 + \cdots + x_n\omega_n$ where $\{\omega_1,\ldots,\omega_n\} = \{1,\sqrt{m}\}$, $\{1,\theta,\theta'\}$, $\{1,\theta,\theta',\sqrt{m},\theta\sqrt{m},\theta'\sqrt{m}\}$ for $n = 2,3,6$ respectively